MATHEMATICS RESEARCH DEVELOPMENTS

STRUCTURAL REALITY

MATHEMATICS RESEARCH DEVELOPMENTS

Additional books in this series can be found on Nova's website
under the Series tab.

Additional E-books in this series can be found on Nova's website
under the e-book tab.

MATHEMATICS RESEARCH DEVELOPMENTS

STRUCTURAL REALITY

MARK BURGIN

nova publishers
New York

For permission to use material from this book please contact us:
Telephone 631-231-7269; Fax 631-231-8175
Web Site: http://www.novapublishers.com

Library of Congress Cataloging-in-Publication Data

Burgin, M. S. (Mark Semenovich)
Structural reality / Mark Burgin.
p. cm.
Includes bibliographical references and index.
ISBN 978-1-62081-641-7 (hardcover)
1. Logic, Symbolic and mathematical. 2. Model categories (Mathematics) 3. Hylomorphism. I. Title.
QA9.7.B87 2011
117--dc23
2012013586

Published by Nova Science Publishers, Inc. † New York

CONTENTS

PREFACE

More and more researchers are beginning to comprehend importance of structures and now the notion of a *structure* is extensively used in a variety of fields. Mathematicians invent and explore mathematical structures, physicists study physical structures, chemists build and investigate chemical structures, biologists work with biological structures, and so on. Actually, all scientists study theoretical structures, assigning these structures to natural phenomena. In such aq way, theoretical structures become models of studied phenomena and scientific laws describe properties and performance of these structures, assuming that these properties reflect attributes and behavior of studied phenomena. Some physicists advocate an extreme approach to physics, according to which the external physical reality is assumed to be a mathematical structure (the Mathematical Universe Hypothesis). At the same time, structures occupy the central place in mathematics. In the beginning, mathematics was the discipline that studied numbers and geometrical shapes. However, the development of mathematics changed its essence and it has become the discipline that studies mathematical structures. For instance, Bourbaki defined pure mathematics as nothing other than the study of pure structures.

Structures are also extremely important in computer science, programming, and computer technology. Data structures form the core of information technology. It is impossible to build a database without explicit construction of data structures. The title "Algorithms + Data Structures = Programs" of the book written by Wirth emphasizes importance of data structures for programming and software development. Computer architecture is a structure of a computer. Structural analysis is used in software technology. Researchers study *knowledge structuration* for the Web. In computer programming, there is a direction called *structured programming*.

Physicists started to understand that they study structures of physical phenomena modeling physical structures by mathematical structures. One of the main goals of chemistry is to reveal the structure of chemical elements and compounds due to importance of these structures. For instance, the spatial structure of atoms, chemical elements, and molecules determines many properties of these chemical systems.

In sociology, there is a theory of *structuration*. There are also a *structural direction* and *structure-nominative approach* in methodology of science. Linguistics studies languages and communication by means of a variety of structures because any developed natural language, such as English or Spanish, comprises a diversity of structures, which allow this language to perform its functions. That is why *structural linguistics* became one of the central fields in linguistics.

In the 20th century, information and knowledge became the prime interest in several areas: in artificial intelligence, science, business, and technology to mention but a few. Understanding of the phenomenon of information is impossible without structures. The new strong emphasis on knowledge explicated importance of the knowledge structure. Different methodologies have been developed to reconstruct the knowledge structure and utilize it for different purposes, such as learning, knowledge management, knowledge representation and knowledge processing.

All this brought forth a general approach in modern science and philosophy called *structuralism*. It is based on revealing intrinsic structures kept invariant relative to transformations, while assuming that different things and phenomena have similar essence if their structures are identical. In this context, to investigate a phenomenon means to find its structure. Structuralism comprises different disciplines with the main emphasis on structure, such as *structural linguistics*, *structuralist economics*, *structural psychology*, and *structural anthropology*. Now one of the most active directions in philosophy is *structural realism*, which asserts that if we can know anything about the outer world and ourselves, then we can know only structures. This book integrates all structural studies into one comprehensive outlook, enlightening the concept of *structure*, unifying different approaches and theories based on structures and providing new tools for structural research. The new approach may be called *neostructuralism* or *multistructuralism*.

This book is aimed at a wide audience of readers who want to better understand the world in which they live. That is why the author tried to write the main body of the book, that is, all its chapters, without mathematical formulas and formal constructions. However, for a mathematician, it is impossible even to imagine such an important topic as structural reality without a mathematical theory. So, formal constructions, exact definitions, mathematical formulas and examples of advanced mathematical structures are given in the Appendix.

The book has eight chapters. In Chapter 1, we analyze the concept of *structure* and its role in cognition and being explaining why *structure* is a basic category. It is clarified why the conventional approach to structures is very limited and has many shortcomings. This was the main reason for the decline of structuralism in the 20th century. The new approach developed in Chapter 1 and formalized in Appendix, allows elimination of these shortcomings, as well as achieving a new understanding of structures and developing a novel typology of structures. This typology helped to discover types of structures unknown before. Chapter 2 gives an exposition of structures in mathematics and logic, exploring such enthusiastically discussed problems as whether all mathematical objects are structures and what objects mathematics studies. In Chapter 3, we contemplate structures in science, medicine, engineering and methodology of science, providing various clarifying examples. In Chapter 4, we describe such important trends in scientific and philosophical thought as *structuralism* and *structural realism*. Chapter 5 gives an exposition of modes and forms of reality. In contrast to the conventional approach where all phenomena and objects are separated into two groups – real and not real, here we show that there are many sorts, modes, forms, dimensions, gradations and types of reality, which form an elaborate structure. This chapter does not contain mathematics but demands mental efforts to comprehend the diversity of the world. Chapter 6 studies structures of such fundamental conceptions as signs, symbols and concepts. Chapter 7 explores the global structures of the world as whole and its reflection in language and human culture. We consider ideas of different philosophers and scientists, which are used for developing the most comprehensive global structure of the world called the Existential Triad. This global structure is reflected in a variety of particular domains, such as language, the human brain and computers. Chapter 8 contains conclusions and direction for further research in the theory of structures. Mathematical formalizations, exact definitions and examples from mathematics, information technology and physics are given in Appendix.

ACKNOWLEDGMENTS

Many wonderful people have made contributions to my efforts with this work. I am especially grateful to Nadya Columbus and Maya Columbus from Nova Science Publishers, for their help in bringing about this publication. I would like to thank the teachers and especially, my thesis advisor, Alexander Gennadievich Kurosh, who helped shape my scientific viewpoint and research style. In developing ideas on the structural aspects of science, mathematics and reality, I have benefited from conversations with many friends and colleagues who have communicated with me on problems of computation, communication and networking. Thus, I am grateful for the interest and helpful discussions with those who have discussed with me these problems. I greatly appreciate advice and help of Andrei Nikolayevich Kolmogorov from Moscow State University in the development of the holistic view on mathematics and its connections with the physical world. Credit for my desire to write this book must go to my academic colleagues. Their questions and queries made significant contribution to my understanding of structures and their importance for science. Collaboration with John Schumann from UCLA extended my understanding of symbolic structures. Collaboration with Yuri Milov from the Kiev Institute of Philosophy was useful for augmenting foundations and broadening applications of the Existential Triad of the World. I would particularly like to thank many fine participants of the Seminar of Foundations of Mathematics and Computer Science at the National Academy of Sciences of Ukraine and of many seminars, conferences and congresses where I presented my results on the structural aspects of science, mathematics and reality for extensive and helpful discussions that gave me much encouragement for further work in this field. I would also like to thank the Departments of Mathematics and Computer Science in the School of Engineering at UCLA for providing space, equipment, and helpful discussions.

Chapter 1

INTRODUCTION: STRUCTURE AS A BASIC CATEGORY

The great Aristotle (384-322 BCE) taught that each thing (physical object) is a *synholon*, which is a composition of *substance* (*hyle*) and *form* (*eidos*) (see (Aristotle, 1984)). Form comes into matter, but neither of them can exist independently in the physical world. Any form is based on a *prote hyle* (a "first matter"), which is the *hypokeimenon* (foundation) of the form. Forms, according to Aristotle, determine essence of things. A philosophical teaching about relations between form and substance is called *hylomorphism* (Wulftange and Greene, 1952; Williams, 1986).

Times changed and scientists began studying nature instead of philosophers. However, scientists came to the same conclusion as Aristotle. In doing this, they used a more exact term *structure* instead of the term *form*, assuming that although the Greek word *form* is usually interpreted as the sight or appearance of a thing, Aristotle treated forms in a more general way. Later findings demonstrated that a form can be conceived as a structure comprehended by senses (Burgin, 2011).

Thus, in accordance with Aristotle, the outstanding physicist Arthur Stanley Eddington (1882-1944) wrote that physical reality has *substance* and *structure* (Eddington, 1922). This statement shows that structures are prevalent in the physical world any physical system has its structure.

In a similar way, another outstanding physicist Erwin Schrödinger (1887-1961) asserted that structure is the basic feature of an object. He wrote (Schrödinger, 1952):

"… in palpable bodies, composed of many atoms, individuality arises out of the *structure* of their composition, out or shape or form, or organization, as we might call it in other cases. The identity of the *material*, if there is any, plays a subordinate role. You may see this particularly well in cases when you speak of 'sameness' though the material has definitively changed."

As we know *substance* is only a general term (name), which has no comprehensive definition because any attempt to further this naming immediately involves attachment of some structure. At the same time, as we will wee, it is possible to develop the general concept of *structure* with a definite extent of precision.

One more great physicist Paul Adrien Maurice Dirac (1902-1984) asserted that with the advance of physics, physical objects start becoming more structures than real things. The contemporary physicist Max Tegmark (2008) advocates an even more extreme approach to physics, where the external physical reality is assumed to be a mathematical structure (the

Mathematical Universe Hypothesis). He believes that this unification of physics and mathematics would allow physicists to understand reality more deeply than most dreamed possible.

Structures appear not only in physics. Personality has its structure, there is the structure of a novel, any language contains many structures and even a dream has its structure. As a result, in the context of the theory of named sets, it was discovered that everything in the world has a structure, even chaos (Burgin, 1997; 2011). Vacuum is considered as absence of matter, void, but physicists study the *structured vacuum* (Mueller and Rafelski, 1985). Thus, structures exist not only in languages, society or human personality, but everywhere. Consequently, it is necessary to study structures not only in linguistics, psychology and anthropology, which were the first areas where the structural approach called *structuralism* was applied, but in all sciences: in natural sciences, such as physics, social sciences and humanities, such as sociology and psychology, and technical sciences, such as computer science.

Researchers have already come to awareness of necessity of structures in their research and now the notion of a *structure* is extensively used in a variety of fields. Mathematicians invent and explore mathematical structures, physicists study physical structures, chemists build and investigate chemical structures, biologists work with biological structures, and engineers build various structures such as buildings, bridges, ships, and so on. Actually, all scientists use symbolic structures in their studies, assigning these structures to natural phenomena. Such structures are called models and scientific laws describe properties and behavior of structures, assuming that these properties and behavior reflect attributes and dynamics of studied phenomena. Moreover, we can see that when it is necessary to learn or to create some system or process, it is done, as a rule, by acquiring of knowledge of the corresponding structure. Structures determine the essence of things.

Many concepts are bound with the term *structure*. In organizations, there are *structural dimensions*. In logic, there are *structural rules*. In computer programming, there is *structured programming* (Dahl, *et al*, 1972). *Structural analysis* is used in software technology (Burgin and Chelovsky, 1982; 1984). In computer science, there are Turing machines with a *structured memory* (Burgin, 2005) and *structural complexity* (Balcazar, et al, 1988). There are *structural equations* in artificial intelligence, decision-making and social theory (Goldberger, 1972; Bollen, 1989). There are also: *structural physics*, *structural mechanics*, *structural engineering*, *structuralist economics*, *structural linguistics*, *structural anthropology*, *structural philosophy*, *structural aesthetics*, *structural chemistry*, *structural geology*, *structural geography*, *structural biology*, and *structural medicine* (see Chapter 3). Structures are everywhere.

If we want to discuss and understand the role and place of structures in the world we are living, we need to know what a structure is. Here we describe and discuss the informal notion of *structure*, allowing the reader to avoid unnecessary mathematical complications. However, the concept of structure in rigorous mathematical terms is described in Appendix because for science, it is necessary to have exact and in many cases, formalized definitions and constructions.

Let us start with some examples of structure definitions. According to McGraw-Hill Encyclopedia of Science and Technology (2005), *structure* is an arrangement of designed components that provides strength and stiffness to a built an artifact such as building, bridge, dam, automobile, airplane, or missile. According to the Dictionary of Business Terms

(Friedman, 2000), *structure* is any constructed improvement to a site, for example, a building, a fence and enclosures, greenhouses, and kiosks. At the same time, *financial structure* has a very different meaning. Namely, *financial structure* is the makeup of the right-hand side of a company's Balance Sheet (Friedman, 2000). In archeology, *structure* refers to the basic framework or form of society. In geography and geology, *structure* is a configuration of rocks of the earth's surface. The *chemical structure* refers both to molecular geometry and to electronic structure. In addition, the *structural formula* of a chemical compound gives a graphical representation of the *molecular structure*, describing the atom arrangement. A *social structure* is a pattern of relations. In mathematical logic, a structure is a logical or mathematical object that gives semantic meaning to expressions of a logical language. The *urban structure* is the arrangement of land use in urban areas, while the *urban spatial structure* concerns the arrangement of public and private space in cities and the degree of connectivity and accessibility (see, for example, (Hoyt, 1939)). In mineralogy and crystallography, *crystal structure* is a unique arrangement of atoms or molecules in a crystalline liquid or solid. A crystal structure is composed of a *pattern*, which is a set of atoms arranged in a particular way, and a *lattice*, which exhibits long-range order and symmetry being a periodic array of *unit cells* formed according to the pattern.

One more definition:

A structure refers to a body or system of connected parts used to support a load.

All these and many other concepts describe only specific kinds of structures. At the same time, there are also general definitions on the intuitive level. For instance, many assume that a structure means a complex of relations comparable with one another in principle. In mathematics, simple relations are represented by pairs of objects between which this relation exists. For instance, the *order relation* on numbers is represented by the set of all pairs of numbers in which the first number is less than the second one, for example, the pair (1, 3) belongs to this relation, while the pair (7, 3) does not belong. More complicated relations involve several objects and are represented by collections of objects between which this relation exists.

However, even in mathematics, there were problems with an exact definition of the concept of structure. For instance, the famous mathematician and philosopher Bertrand Arthur William Russell (1872-1970) had a vague and informal understanding of structure in spite of being one of the first researchers who brought this concept to the forefront of science, mathematics and philosophy. In his works, Russell used the concepts *structure* and *relation-number* interchangeably, talking of the structure of a relation or of a system of relations defined over a given domain. According to him, the structure of a relation is "the class of relations similar to the given relation" (Russell, 1927). Russell employs the notion of *similar relations*, which may be interpreted as *isomorphic relations*, to convey the idea that only the logico-mathematical properties of a relation are important and not the relation itself because there is no direct access to it.

Often structures are considered as abstract representations of systems. For instance, the contemporary philosopher Stewart Shapiro (1997) describes a structure in the following way:

"I define a system to be a collection of objects with certain relations. … A structure is the abstract form of a system, highlighting the interrelationships among the objects, and ignoring any features of them that do not affect how they relate to other objects in the system."

Later Shapiro (2010) simplified this definition:

"A *structure* is the abstract form of a system, which ignores or abstracts away from any features of the objects that do not bear on the relations."

In a similar way, by the term *structure* (*Aufbau*), a prominent philosopher Rudolf Carnap (1891-1970) refers to all the formal properties of an object or relation (Carnap, 1928). The formal properties of an object (or of a concept, statement, or relation) are those properties which define its logical structure, but not its material content. Scientific statements may describe the structural properties of objects or relations, and may take the form of either property descriptions or relation descriptions.

The outstanding linguist Louis Hjelmslev (1899–1965), one of the protagonists of structural linguistics, considered a structure to be an autonomous essence with inner relations (Hjelmslev, 1958).

In their study of structural realism (see Chapter 4), Frigg and Votsis (2011) use the following definition:

"A *structure* S consists of (a) a non-empty set U of objects, which form the domain of the structure, and (b) a non-empty indexed set R (that is, an ordered list) of relations on U, where R can also contain one-place relations (that is, monadic properties)."

Here a *one-place relation* is a relation between elements of the structure and some other sets.

In a more general context, it is assumed that a *structure* is, or is a representation of, a complex entity that consists of parts in relations to each other (Grossmann, 1990). Note that a system can have many such representations and thus, many structures.

There are also strict mathematical definitions of structures, which are considered in Appendix. Definitions from one class of them are constructed by means of set theory (Bourbaki, 1957; 1960; Robinson, 1963; Yaglom, 1980; Tegmark, 2008), definitions from another class use mathematical categories (Bucur and Deleanu, 1968), definitions from the third class use formal schemas (Burgin, 2010), the fourth class is developed in the context of such a field as mereology (Burgin, 2011), and there are characterizations of structures by direct axioms, that is, anything that satisfies such axioms constitutes a structure (Shapiro, 1997).

The most popular definition assumes that a structure is a set of elements with various relations between them. In the theory of structures originated in (Burgin, 2010; 2011), such structures have the name *structures of the first level* because they are the simplest kind of structures. Exact definitions of structures of the first and higher levels are given in the Appendix for mathematically oriented readers.

Structures of the first level are easily comprehensible and have transparent graphical or visual representations, for example, as diagrams, flowcharts, graphical schemas and nets, as well as formalized symbolic representations, for example, as formulas, equations, algorithms, logical propositions and other mathematical constructions. For instance, when a structure includes only relations between pairs of elements, it is portrayed by nodes/points, which represent parts or elements of the whole and are connected by edges, which represent relations between elements. In mathematics, such a system is called a graph. Figure 1.1 provides several examples of graphs.

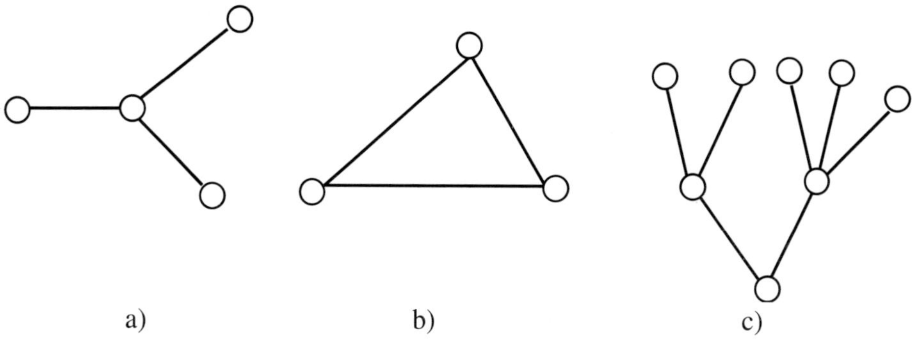

Figure 1.1. Graphical representations of structures in the form of a graph: graph (a) is called a binary tree; graph (b) is a triangle; and graph (c) is called a tree.

As visual representations alleviate comprehension and cognition, we use many diagrams and graphical schemas in this book to make our exposition more transparent and to achieve better understanding. We call such representations by the name *visual structures*. It is necessary to understand that often visual structures give a very approximate image of real phenomena. To have sufficiently precise structural representations, scientists use mathematics with its symbolic structures.

There are many examples and definitions of structures but in spite of structure utilization in a variety of areas and many achievements of structuralism as a general methodology and philosophy of the structural approach, different meanings are assigned to the word *structure* in different areas and there is no comprehensive definition of *structure* accepted by all researchers. Researchers mostly study structures as they are interpreted in mathematics, natural sciences, linguistics, social sciences, psychology, etc. (Piaget, 1971; Caws, 1988). Resnik (1997) even asked a question whether it is possible to give an exact definition by building a comprehensive mathematical model for the notion of structure.

So, to build such a model, it is necessary to better understand what interpretations the notion of structure has. When people speak about a structure, it is usually the structure of something, i.e., of some object, system or entity, for example, the structure of the universe. This shows that there is a duality between objects/entities and structures. That is, having an object, for example, a molecule or organization, people look for its structure and this structure gives knowledge about the object/entity.

Note that that usually it is assumed that an object has only one structure (if any). That is why mathematicians use such expressions as "the structure of a triangle" or "the structure of the triangle *ABC* from Figure 1.2."

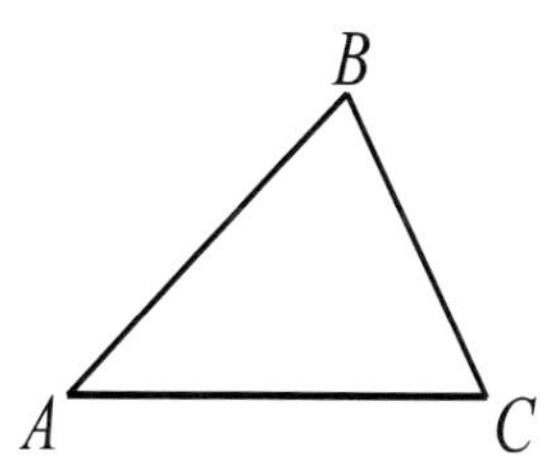

Figure 1.2. A triangle *ABC*.

In a similar way, physicists write about the structure of an atom or the structure of the Solar System. However, virtually any system has several structures. Even such a simple system as a triangle has, at least, three structures:

Structure 1: Three vertices connected by segments of a straight line, that is, in this case, vertices are elements of the structure and there is one relation "connected by a segment of a straight line".

Structure 2: Three segments of a straight line such that each two of them have a common endpoint, while all three do not have common endpoints, that is, in this case, segments are elements of the structure and there is one binary relation "have a common endpoint".

Structure 3: Three vertices and three segments of a straight line such that each vertex coincides with the endpoints of exactly two segments, that is, in this case, segments elements of the structure are vertices and segments and there is one binary relation "to coincide with an endpoint".

Although these structures are, in some sense, equivalent, they are different as structures and as systems. At the same time, these three structures/systems represent the same shape.

The situation when a structure is derived from or corresponded to an object/entity is presented by the following triad.

$$\textbf{objects/entities} \xrightarrow{\text{derived from}} \textbf{structures} \tag{1.1}$$

However, it is possible to go into the opposite direction, that is, to assign objects/entities to structures. For instance, taking a structure Q, we can (at least, theoretically) correspond all objects/entities that have this structure Q to Q. We can also ask whether an object has the given structure Q. These situations invert the triad (1.1) and we obtain the following triad.

$$\textbf{structures} \xrightarrow{\text{assigned to}} \textbf{objects/entities} \tag{1.2}$$

Besides, an object can be a structure itself, for example, a mathematical structure. This situation does not change Diagrams (1.1.) and (1.2) because it is possible to consider a structure of a structure. For instance, we can take such an object as the set $\boldsymbol{N}$ of natural numbers, that is, numbers 1, 2, 3, and so on. Many mathematicians assume that this is a structure. At the same time, this object $\boldsymbol{N}$ has an order structure, in which there are order relations between numbers indicating when a number is larger or smaller than another number. In addition, this object has the algebraic structure determined by addition of numbers.

We can observe a multitude of structures around us but an interesting fact is that the most fundamental structure was discovered. Namely, it is demonstrated in (Burgin, 2011) that all structures are built of this fundamental structure. It is called a *fundamental triad* or a *named set.*

A fundamental triad (named set) has the following graphic representations.

$$\textbf{Entity 1} \xrightarrow{\textbf{connection}} \textbf{Entity 2} \tag{1.3}$$

or

$$\textbf{Essence 1} \xrightarrow{\textbf{correspondence}} \textbf{Essence 2} \tag{1.4}$$

We can see a variety of fundamental triads around us. When two people are speaking, they form a fundamental triad. Indeed, we have Individual 1 and Individual 2 who are connected by their speech (pronounced words). When two people are speaking by conventional phones, they are connected by wires. When two people are speaking by cell phones, they are connected by radio waves. In all cases, they form a fundamental triad.

When it is raining, we see the fundamental triad that consists of a cloud(s) (Entity 1), the Earth where we stand (Entity 2) and flows of water (a correspondence). When we see a lightning, we see another fundamental triad that consists of a cloud(s) (Entity 1), the Earth where we stand (Entity 2) and the lightning (a correspondence).

People often meet fundamental triads in their everyday life and find them in many fundamental structures of the universe. People and their names constitute a fundamental triad/named set. Cars and their owners constitute another fundamental triad. Books and their authors constitute one more fundamental triad. A different example of a fundamental triad is given by the traditional scheme of communication:

$$\textbf{Sender} \longrightarrow \textbf{Receiver}$$

In this case, the connection may be one of the following: a channel, the communication media, or a message.

Analyzing the structure of a fundamental triad (named set) **X**, we see that it consists of three components, that is, $\mathbf{X} = (X, f, I)$. Each component of a fundamental triad plays its unique role and has a specific name. Reflecting analytic role of the structure *fundamental triad*, we call X the *support*, I the *reflector*, and f the *reflection* of the fundamental triad. In the same structure $\mathbf{X} = (X, f, I)$ considered as a named set, X (that is, Entity 1 or Essence 1) is called the *support*, I (that is, Entity 2 or Essence 2) is called the *component of names* or *set of names*, and f is called the *naming correspondence* of the named set **X**. Thus, informally a fundamental triad may be described by the expression:

$$f \text{ connects } X \text{ with } I \tag{1.5}$$

or

A fundamental triad is a unity of three elements where the second element is a connection/relation between the first element and the third element.

For instance, taking a group of people and their names, we see that it is a named set (fundamental triad), in which the group of people forms the support, the set of their names is the component of names (reflector) and connections between people and their names mold the naming correspondence (reflection).

Utilization of two names, *named set* and *fundamental triad*, for the same structure implies difference in perspective and intuition. The name *fundamental triad* reflects unity of the components, wholeness, and elementarity of the whole triad. In particular, some fundamental triads are elementary being actually indivisible (Burgin, 2011). The name *named set* points to a tentative inner structure and compositionality of the named set components. The main idea formalized in the concept of a named set is: *elements of a set (actually, any objects with which people are dealing) have names*. Thus, the model example of a named set is a set elements of which have names, in particular, a group of people and their names.

Explicating connections between different entities and a possibility to have connections of diverse types, named sets (fundamental triads) reflect and represent basic aspects of reality. As Mills (2004) writes, "to understand anything, whether it is the operation of a complicated mechanism or the complex social factors that underlie almost any human situation, understanding it means seeing the connections." Named sets (fundamental triads) portray connections in an explicit (often visible) form as the basic component of the whole structure.

Fundamental triads (named sets), as the most encompassing and fundamental mathematical construction, encompass all generalizations of ordinary sets and provide unified foundations for the whole mathematics (Burgin, 2004; 2011). The concept of a named set (fundamental triad) was discovered in 1980s in an attempt to extract the common essence of those mathematical constructions that are more general than sets (fuzzy sets, multisets, *L*-fuzzy sets, rough sets, genuine sets, etc.).

There are different sorts of fundamental triads (named sets): set-theoretical fundamental triads, categorical fundamental triads, algorithmic fundamental triads, mereological fundamental triads, etc. The most routine fundamental triads are set-theoretical ones because all mathematicians are accustomed to talk set-theoretical language and the whole mathematics is developed using set-theoretical fundamental triads (named sets) as its base. They are called *set-theoretical named sets* because all their components are sets. A visual representation of set-theoretical triads (named sets) is given in Figure 1.3.

At the same time, there are many named sets that are not set-theoretical. For instance, an *algorithmic named set* $\mathbf{A} = (X, R, Y)$ consists of the set X of inputs, the set Y of outputs and rules R of transformation of the input into the output. Such rules are called algorithms when they are understandable and simple for realization. We know that there are many systems of rules that represent in a computational way one and the same function, that is, a given input x may be transformed in the same output y by different rules. This means that one set-theoretical named set is related to many algorithmic named sets, which define the same function. Thus, algorithmic named sets are different from set-theoretical named sets.

Many mathematical systems are particular cases of fundamental triads. The most important of such systems are relations, functions, as well as fuzzy sets (Zadeh, 1965) and multisets (Knuth, 1997) as natural generalizations of the concept of a set and have many important applications. It is interesting that functions and binary relations, that is, relations between pairs of objects, have been represented in the form of fundamental triads even before the concept of a fundamental triad was discovered (see, for example, (Bourbaki, 1960; Germain, 1971)). Other systems, such as graphs and hypergraphs (Berge, 1973), Boolean valued sets (Bell, 2005), topological and fiber bundles (Husemöller, 1994), and many other mathematical structures also are fundamental triads or are constructed from fundamental triads as construction components. Moreover, any ordinary set is, as a matter of fact, some named set, and namely, a singlenamed set, that is, such a named set (fundamental triad) in which all elements have the same name (Burgin, 2011).

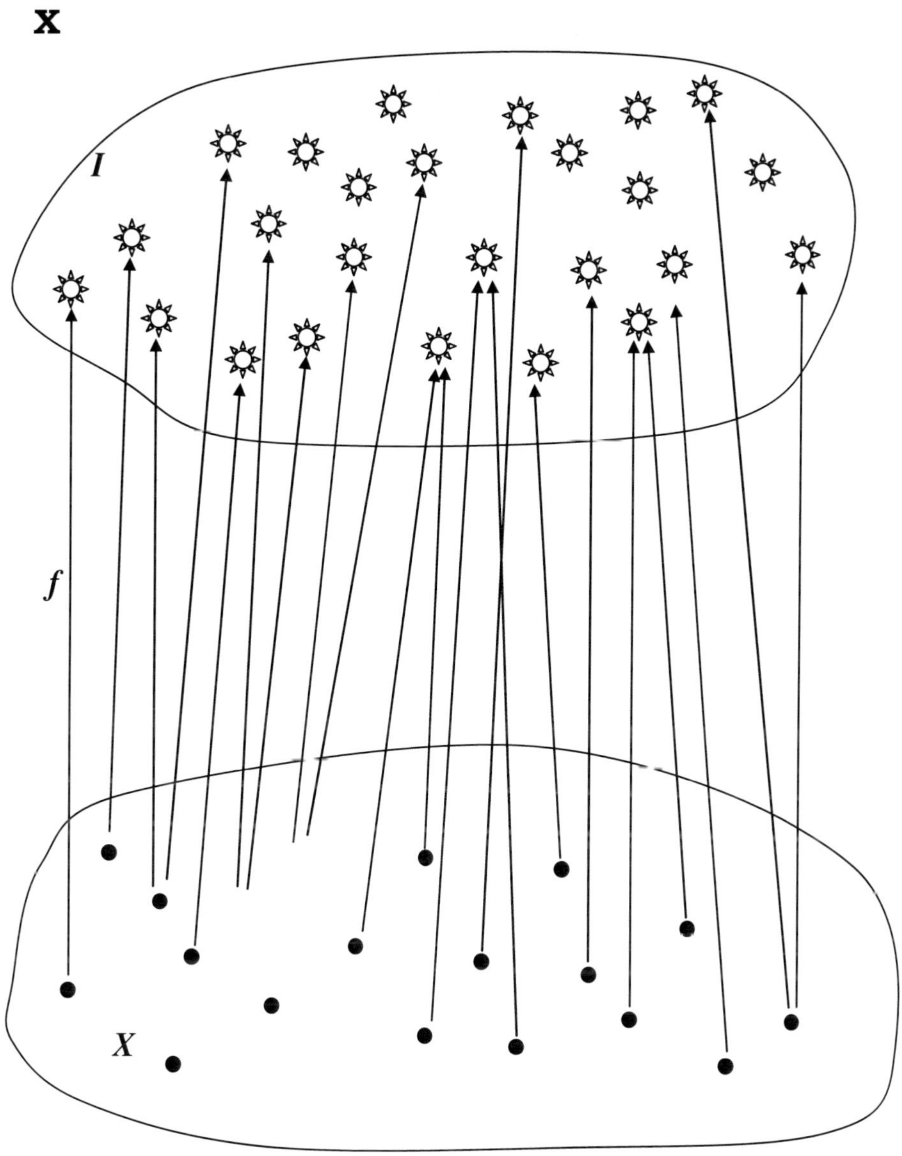

Figure 1.3. A set-theoretical named set X = (*X*, *f*, *I*)

Indeed, any set that is considered either in mathematics or in real life has a name. As Poincare (1908) wrote, without a name, no object exists in science or mathematics. The name of a set forms this set from separate elements by attaching to each of them a common name. Namely, all elements from a set *X* have the common name "an element from the set *X*." As a result, any set implicitly exists as a singlenamed set. In conventional set theories, supports of these singlenamed sets are considered as autonomous objects, giving the abstraction of ordinary sets. Thus, when we write that a set X is equal to a set Y, it means that these singlenamed sets have the same support, and in many cases, it is possible to ignore existence of the element names.

Thus, if we have a set A, we naturally represent it by the siglenamed set in which all elements have the common name "*an element of the set with the name A*". It is interesting that to build a more flexible and transparent set theory in the classical sense, some authors use named sets instead of ordinary sets (see, for example, (Holmes, 1998; Jones, 2007)). It is interesting that utilization of singlenamed sets instead of ordinary sets allows one to solve some problems in science, for example, Gregg's paradox in biology (see (Ruse, 1973; Burgin, 1983)).

Many basic structures outside mathematics are also fundamental triads. For instance, explicating the structure of concepts, proper names and propositions in the theories of Russell and Frege, Tatievskaia (1999) demonstrates that all of them have the form of a fundamental triad (see Chapter 6).

When researchers study information and information processes, fundamental triads become extremely important. Each direction in information theory has fundamental concepts that are fundamental triads and models of which are also either fundamental triads or systems built of fundamental triads (Burgin, 2010). For instance, the relation between information and the receiver/recipient is a fundamental triad.

Note that we discern triads and triples. Any three objects form a triple, while to constitute a triad, it is necessary that some connection or relation exists between such three objects.

There are also other basic triads. One of them is a *bidirectional triad*, which has the following graphic representations.

$$\textbf{Entity 1} \xleftrightarrow{\textbf{connection/relation}} \textbf{Entity 2} \qquad (1.6)$$

or

$$\textbf{Essence 1} \xleftrightarrow{\textbf{correspondence}} \textbf{Essence 2} \qquad (1.7)$$

Similar to the fundamental triad, a bidirectional triad $\mathbf{X}$ consists of three components, that is, $\mathbf{X} = (X, f, I)$.

Any bidirectional triad is the *coupling* of two fundamental triads. For instance, the bidirectional triad (1.6) is the *coupling* of two fundamental triads (1.8) and (1.9).

$$\textbf{Entity 1} \xrightarrow{\textbf{connection/relation 1}} \textbf{Entity 2} \qquad (1.8)$$

and

$$\textbf{Entity 2} \xrightarrow{\textbf{connection/relation 2}} \textbf{Entity 1} \qquad (1.9)$$

One more basic triad is a *triangular triad* also called a *system triad* (Barantsev, 1982; 1989), which has the following graphic representation.

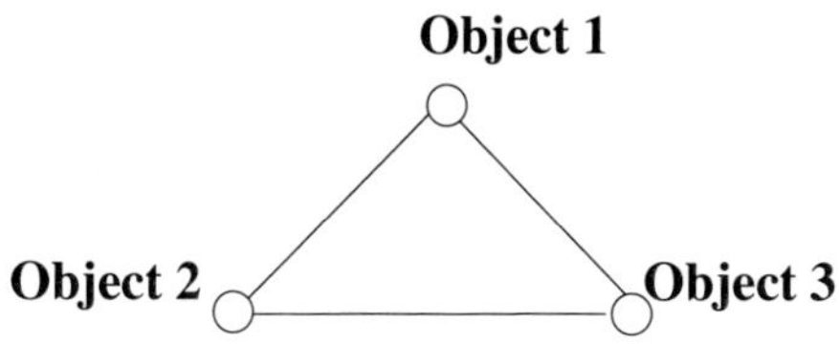

Figure 1.4. A triangular or system triad.

Triangular/system triads have been featured through several millennia of the Earth civilization, starting with myths and legends of ancient people. Many examples of triangular triads are described in Chapter 7.

A well-known example of system triads is the famous *dialectic triad*

(*thesis, antithesis, synthesis*)

Its graphical representation is given in Figure 1.5 and its components are usually described in the following way:

- The *thesis* is an intellectual proposition.
- The *antithesis* is the negation of the thesis as a reaction to the proposition.
- The *synthesis* solves the conflict between the thesis and antithesis by reconciling their common truths, and forming a new proposition.

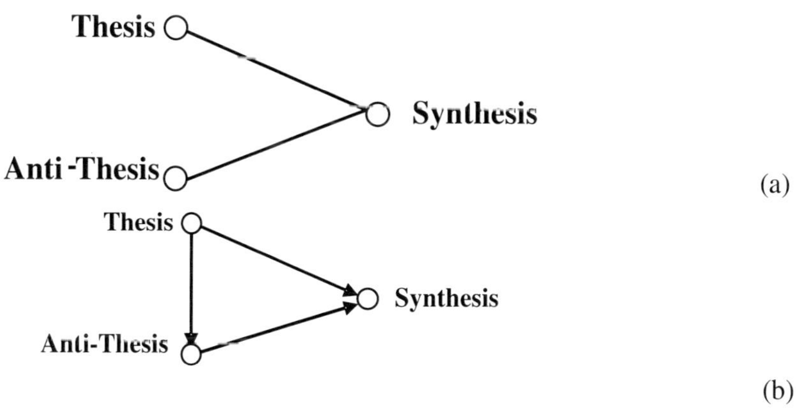

Figure 1.5. The dialectic triad in the (a) disjoint form, and (b) unified form.

The dialectic triad is often attributed to the outstanding German philosopher Georg Wilhelm Friedrich Hegel (1770 –1831) and called the *Hegelian dialectic triad*. However, according to some of his biographers, Hegel never used the term himself. According to other biographers, Hegel used this structure only once, attributing the terminology to another German philosopher Immanuel Kant (1724-1804). However, according to the majority of sources, the dialectic triad as a basic knowledge acquisition process was explicitly developed by Johann Gottlieb Fichte (1762 - 1814) in his book *Introduction to the Wissenschaftslehre*.

At the same time, as Wyatt writes (http://www.tameri.com/csw/exist/hegel.shtml), it was Hegel's contention that the Absolute Truth was a collection of all possible triads in the universe. Mankind could discover truth only if people could recognize every phenomenal triad that existed.

It is interesting to know that *triad* as the basic structure was also very popular in works of Neoplatonic philosophers, such as Plotinus (ca. 204–270), Iamblichus (c. 250-325) and Proclus (ca. 410–485).

One more triangular or *codialectic* (that is, dual to dialectic) *triad* is basic for Chinese philosophy. The indispensable conception of Chinese philosophy is *Tao* or *Dao*, which is translated from Chinese as *way*, *path*, *route*, or sometimes more loosely, as *doctrine* or *principle*, and signifies the primordial essence and fundamental nature of the universe.

This metaphysical concept originated with the great Chinese philosopher Laotzu, also Laotse or Laozi (ca. 604-517 BCE). In the *Tao Te Ching*, Laotzu explains that Tao is not a *name* for a *thing* but the underlying natural order of the universe whose ultimate essence is difficult to circumscribe (Dao De Jing-32. Laotzu). As a result, Tao is "eternally nameless" and must to be distinguished from the countless "named" things which are considered to be its manifestations. The concept of Tao was later adopted in Confucianism, Chán and Zen Buddhism and more broadly throughout East Asian philosophy and religion in general.

Two other basic conceptions of Chinese philosophy are *Yin* and *Yang*, which are the two primal cosmic forces in the universe. *Yin* (the Moon) is the receptive, passive, dark, cold female force, impersonating stillness. *Yang* (the Sun) is the bold, active, light, hot masculine force, impersonating movement. However, in contrast to Western philosophy, these oppositions do not contradict one another. The *Yin-Yang principle* represents the philosophy of balance, where two opposites co-exist in harmony and are able to transmute into each other. This opposition and combination of the universe's two basic principles of Yin and Yang is a large part of Chinese philosophy, and is an important feature of Taoism, both as a philosophy and as a religion. The Yin-Yang principle is represented in Taoism by the Taijitu, which literally means the "*diagram of the supreme ultimate*".

Figure 1.6. Taijitu diagram.

Taijitu brings us to the *dialectic triad* of Taoism.

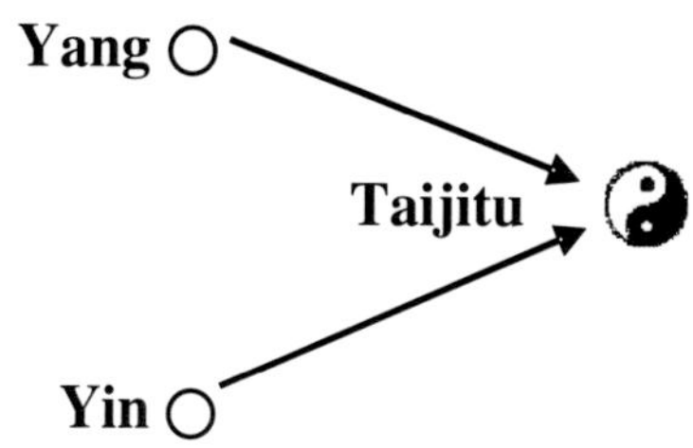

Figure 1.7. The dialectic triad of Taoism.

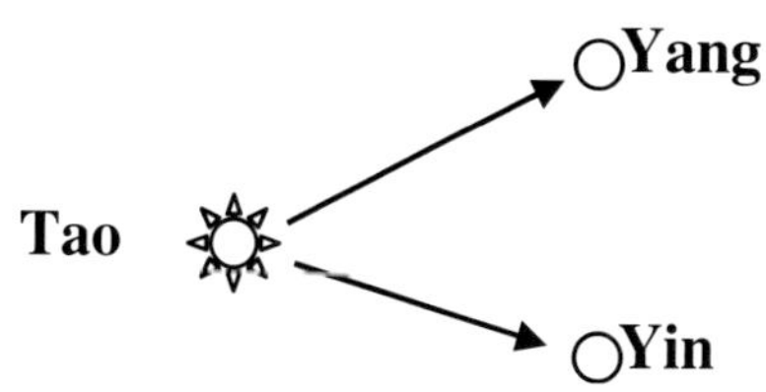

Figure 1.8. The *codialectic triad* of Tao - Yin – Yang.

Yin and Yang, as well as all other opposites, are manifestations of the single Tao, and are therefore not independent from one another, but rather a variation of the same unifying force throughout all of nature. This is reflected in Figure 1.7.

Speaking about structures, we see that there are pure or abstract symbolic structures, which we encounter, for example, in mathematics, and there are structures of some systems, which we will call embedded structures.

Working with structures, it is also important to discern pattern structures and individual structures.

A *pattern structure* is a structure common to many systems. For instance, structure of a triangle is the same for a multitude of triangles.

An *individual structure* is a structure of some individual object. For instance, taking a definite triangle and determining its structure so that it would define a unique triangle, we obtain an individual structure of this triangle.

Objects with a structure *R* are called *instantiations* of *R*. Thus, a pattern structure has many instantiations, while an individual structure has only one instantiation.

Some researchers (see, for example, (Bourbaki, 1960; Resnik, 1981; 1999; Shapiro, 1989; 1997)) deal only with pattern structures ignoring individual structures because they assume that mathematics studies patterns. However, in principle, mathematics can also study individual structures although any mathematical structure can be and often is instantiated by various objects that have this structure.

Whether a structure is individual or pattern often depends on the class of considered systems. Individual structure is like an individual (proper) name. Indeed, in a small group, people's names are usually individual – different people have different names. However, in big groups, we can often find different5 people with the same name. Likewise, going to a larger realm, it is often possible to find systems with the same structure.

Now we proceed to the advanced understanding of structures, which demands more mental efforts. So, the reader may skip the end of this chapter in the first reading and can directly go to Chapter 2.

In the process of exploration, the traditional approach to structures has been criticized as too simplified and incomplete. For instance, based on this understanding, Carter (2008) argues that a set taken by itself may not be regarded as a structure. Even more, she contrasts sets and structures, demonstrating that the relation that exists between the structure and the set on which this structure is defined (as a system of relations) is important for mathematical practice. On similar grounds, Parsons (1990) demonstrates that mathematical objects include relations not only to objects from the same set but also relations to other objects, which Parsons call external relations. Consequently, existing definitions do not represent the intuitive concept of structure as the organization of parts into the whole.

This shows that it is necessary to have a more general concept of structure. Such a concept was developed in (Burgin, 2010; 2011). It eliminates all objections to the structural understanding of mathematical objects. Objections of Parsons lose their validity because structures in the more advanced sense include external relations, while the argument of Carter becomes inapplicable because any set has the structure of a named set as it is explained above. In addition (see Appendix), any set has a mereological structure, which is also a mereological named set (Burgin, 2011). Here we give only an informal outline of some necessary extensions of the concept of a structure. It is possible to find rigorous definitions and related mathematical constructions in Appendix.

To achieve our goal in an adequate development of the structure concept, it is useful to discern and take into account four classes of structures:

- An *inner*, or *internal*, *structure* of a system, object or process.
- An *outer structure* of a system, object or process.
- An *intermediate*, or *connection*, *structure* of a system, object or process.
- The *external structure* of a system, object or process.

Note that a system can have several structures, as well as several inner, outer and intermediate structures.

An *inner structure* of a system represents this system in a symbolic form as formed from parts (elements) and connections or relations between them.

An *outer structure* of a system R is the inner structure of a supersystem Q that contains R as an element (part).

Note that the concept of an outer structure of a system depends on the context or environment in which this system is considered. For instance, the outer structure of an individual in his family is different from the outer structure of this individual in society.

An *intermediate structure* of a system R represents in a symbolic form connections, relations and ties between R and other systems, that is, the system R is considered as a whole having (specified) relations with and connections to other systems.

Let us take such a system as a human being A and consider some inner, outer and intermediate structures of A. One of the inner structures of A is the structure Q of the body of A. This structure has such elements/parts as the head, two arms, two legs, and torso. Relations between them show their connections, namely, the head, arms and legs are connected to the torso. Connections of the arms and legs to the torso are direct, while connection of the head to the torso is substantiated by the connecting link – the neck.

An example of an intermediate structure of *A* is the system of properties of *A*, such as the weight, height, age, place of living, profession, etc. Note that any property of an object is a relation between this object and the scale of the property. Thus, a property is always an intermediate relation. Another example of an intermediate structure of *A* is the structure of the attitudes of *A* to other people.

An example of an outer structure of *A* is the structure of family of *A*. One more example of an outer structure of *A* is the structure of the building where *A* lives.

Another example of a structure of *A*, which is usually treated as an inner structure, is the structure of personality considered in Section 3.8.

The *external structure* of a system represents this system in a symbolic form as formed from elements that belong to this system, connections/relations between these elements and elements from other systems, and relations between relations.

Properties of parts/elements are particular cases of such external relations. Usually properties of parts/elements are not included in the conventional structure of a system although Carnap (1928) wrote that a structure consists of formal properties.

Examples of properties of people are their weights or heights.

Note that Frigg and Votsis (2011) also include properties as one-place relations in the definition of a structure. Moreover, they consider the whole hierarchy of properties: properties of elements, properties of properties of elements, properties of properties of properties of elements, and so on.

Thus, we come to a conclusion that a structure is, or is a symbolic representation of, a complex entity that consists of:

- Parts/elements of this entity in relations to each other and other entities, as well as properties of these parts/elements. Relations between parts/elements are called relations of the first level. A well-known example of a mathematical relation is the relation "to be larger than" for natural numbers. For instance, 3 is larger than 1. This is a relation of the first level.
- Relations of these relations to each other, to parts/elements of this entity and to other entities. These relations are called relations of the second level.
- Relations of these relations to each other, to relations considered on the previous level, to parts/elements of this entity and to other entities. These relations are called relations of the third level.

and so on.

Let us consider some examples of relations between relations.

A. The relation "to be brothers" implies the relation "to be relatives".
B. The relation "to live in the same apartment" implies the relation "to live in the same building".
C. The relation "to be younger" is inverse, that is, in some sense, opposite, to the relation "to be older".
D. The relation/property "to be a bachelor" excludes the relation "to be married".

To demonstrate mathematical relations of the higher level, that is, a relation between relations, we consider relations "to be larger than" and "to be larger than or equal to" for natural numbers. For instance, 3 is larger than or equal to 1, and 3 is also larger than or equal to 3. Then it is easy to see that the relation "to be larger than" implies the relation "to be larger than or equal to". It means that if one number is larger than another number, then the first number is larger than or equal to the second number. This implication is a relation between relations and is a relation of the second level.

In the following chapters, we use some mathematical terms ascribing to them not a strict mathematical interpretation but an informal meaning for easier understanding. For instance, comparing two structures, we see (show) that in some cases they are similar. When we have a close similarity, we say that these structures are *homologues*. When there is a natural one-to-one correspondence between parts (components or elements) of these structures and between relations that connect these parts, we say that these structures are *isomorphic*.

Chapter 2

STRUCTURES IN MATHEMATICS AND LOGIC: WHAT MATHEMATICS IS ABOUT

At the beginning, mathematics was the discipline that studied numbers and geometrical shapes/forms (Burton, 1997). However, the development of mathematics changed its essence and mathematics has become the discipline that studies exclusively mathematical structures (Lautman, 1938; Bourbaki, 1948; 1957; 1960; Burgin, 1998; 2011). Numbers and geometrical shapes/forms are still mathematical objects actively studied by mathematicians but they are treated as definite kinds of structures. For instance, according to Shapiro (1997), "the essence of a natural number is the relations it has with other natural numbers."

As Steen (1988) writes, the rapid growth of computing and applications has helped cross-fertilize the mathematical sciences, yielding an unprecedented abundance of new methods, theories, and models. Examples from statistical science, core mathematics, and applied mathematics illustrate these changes, which have both broadened and enriched the relation between mathematics and science. No longer just the study of *number* and *space*, mathematical science has become the science of patterns, with theory built on relations among patterns and on applications derived from the fit between a pattern and observation.

Formal structures occupy the central place in the unified picture of mathematics. For instance, Bourbaki defined pure mathematics as nothing other than the study of pure structure. Dieudonné described the unifying role of the structures in the following way:

"Each structure carries with it its own language, freighted with special intuitive references derived from the theories from which the axiomatic analysis ... has derived the structure. And, for the research worker who suddenly discovers this structure in the phenomena which he is studying, it is like a sudden modulation which orients at once the stroke in an unexpected direction in the intuitive course of his thought and which illumines with a new light the mathematical landscape in which he is moving about.... Mathematics has less than ever been reduced to a purely mechanical game of isolated formulas; more than ever does intuition dominate in the genesis of discoveries. But henceforth, it possesses the powerful tools furnished by the theory of the great types of structures; in a single view, it sweeps over immense domains, now unified by the axiomatic method, but which were formerly in a completely chaotic state."

Some philosophers and many mathematicians also stated that all mathematical objects are structures of some type. For instance, Resnick (1997) explains:

"In mathematics … we do not have objects with an 'internal' composition arranged in structures, we have only structures. The objects of mathematics … are structureless points or positions in structures. As positions in structures, they have no identity or features outside of a structure."

At the same time, not all mathematicians assume that all mathematical objects, and in particular, the set of natural numbers, are structures.

Because the goal of this book is to describe the world of structures not only to people with advanced mathematical knowledge, here we explain the essence and nature of mathematical structures using simple examples.

Let us consider natural numbers 1, 2, 3, 4, 5, … First, we can take the set $\boldsymbol{N}$ of all natural numbers. Doing this, we obtain the basic structure of a set defined by the membership relation. This is a relation between elements and the set. Namely, this relation R_N consists of all pairs $(n, \boldsymbol{N})$ where n is a natural number. In mathematics, we write $n \in \boldsymbol{N}$ where $\in$ is a special symbol invented by mathematicians to denote membership.

In addition, natural numbers are related by means of different operations, such as addition or multiplication. For instance, $1 + 2 = 3$ or $7 \times 5 = 35$. These operations form an algebraic structure on the set $\boldsymbol{N}$.

One more structure on the set $\boldsymbol{N}$ determines order between natural numbers. For instance, 1 is less than 2, while 7 is larger than 5.

One more basic structure on the set $\boldsymbol{N}$ is defined by distances (called metric relations in mathematics) between natural numbers. For instance, it is possible to say that the distance between 1 and 11 is 10, while the distance between 7 and 5 is 2.

As North (2009) writes, the idea of mathematical structure is relatively straightforward. There are number structures, geometric structures, topological structures, algebraic structures, and so forth. Mathematical structure tells us how simpler mathematical objects are organized to form more complex mathematical objects. It is possible to find exact definitions of mathematical structures given by different authors in Appendix.

Mathematicians more and more explicitly use the concept of structure building mathematical theories and more precisely structuring existing concepts. Here are some examples that demand more mathematical knowledge and can be skipped in the first reading. Rocchi (2003) developed a new structural approach to the theory of probability. Explaining that probability theory from the very beginning and onwards has focused on events as sets of outcomes or results, and probability as a measure attached to these outcomes, Rocchi includes the structure of the processes which lead to the outcomes in this picture. This allows him taking into account the details of the processes to achieve a more fundamental understanding of the nature of probability.

Another example of structuration in mathematics is related to concept of a variable. Usually variables in mathematics are identified with letters, that is, with objects that have only outer structures. For instance, in (Stewart, *et al*, 2004; 2009), we can find the following definition.

A variable is a letter that can represent any number from a given set of numbers.

However, utilization of variables in information technology brought forth their inner structure (Burgin, 2011). A *variable* is a fundamental triad of the form

(*name*, interpretation mapping, *domain*)

Components of this fundamental triad are:

- *name* means the name of the variable, for example, the letter x is used in mathematics as a name of an unknown variable, the letter m is used in physics as a name of the variable *mass*, the letter t is used in physics as a name of the variable *time*, the letter T is used in physics as a name of the variable *temperature*, and the letter C is used in economics as a name of the variable *cost*;
- the *domain*, also called *object domain*, of the variable consists of all tentative values of this variable, for example, the domain of the variable m is the set of all non-negative real numbers;
- the *interpretation mapping* is the relation that connects the name of the variable with all elements (objects) from the domain of this variable.

Besides, *structured variables* were introduced and utilized. They have the name that includes names of other variables, for example, structured variables V(x, y), $R(A, B)$, f_t, and $x + y$ include names of variables x, y, t, A and B. Variables names of which are included in the name of a structured variable X are called components of X.

This ends our mathematical examples in this chapter and we are back to the layman level.

An emphasis on mathematical structures had a marked influence *outside* mathematics. For instance, Bourbaki introduced three basic types of structures (algebraic structures, order structures and topological structures) and Piaget (1964) found a close correspondence between these types and the first operations through which the child interacts with the world.

However, different researchers criticized the traditional approach to mathematical structures as too primitive. For instance, based on the conventional understanding of *structure*, Carter (2008) argues that a set taken by itself may not be regarded as a structure. In addition, she demonstrates that the relation that exists between the structure and the set on which this structure is defined (as a system of relations) is important for mathematical practice. In his critic of the structuralist view of mathematical objects (see, for example, (Bourbaki, 1948; 1957; Resnick, 1997; 1999)), which asserts that all mathematical objects are structures *per se*, Parsons (1990) explains that mathematical objects include relations not only to objects from the same set but also relations to other objects, which Parsons call external relations. Thus, only the more general concept of structure developed in (Burgin, 2010; 2011) eliminates all objections to structural understanding of mathematical objects.

Besides, it is interesting to know that in mathematics there are situations when objects with different structures are identified, that is, treated as one and the same object. For instance, the natural number 3 is considered the same as the rational number 3. However, the outer structure of the natural number 3 is the structure of the object $\boldsymbol{N}$ of all natural numbers, while the outer structure of the rational number 3 is the structure of the object $\boldsymbol{Q}$ of all rational numbers. The inner structure of the natural number 3 is the named set (X, f, 3) where X consists of (all) sets that have three elements and f connects all these sets to the symbol 3, while the inner structure of the rational number 3 is the named set (Z, r, 3) where Z consists of (all) fractions 3/1, 6/2, 9/3, … and f connects all these fractions to the symbol 3. This shows that mathematical objects are not structures, at least, in that oversimplified sense that is traditionally assigned to the concept of structure.

It is natural to ask a question why different structures are treated in mathematics as the same object. We can answer this question explaining that mathematics is a tool (mechanism) for simplification through a rigorous generalization and unification. Thus, when the rational number 3, the whole number 3, the integer number 3, the rational number 3 and the real number 3 are comprehended as one and the same object, it provides a useful unification, simplifying many cases of mathematical reasoning.

Now it is possible to answer the question whether mathematics is a science of structures and only of structures. As we already know, Bourbaki (1957; 1960) asserted that mathematics is a discipline that studies mathematical structures and only such structures. This point of view is shared by many mathematicians, philosophers of mathematics (see, for example, (Shapiro, 1997; Resnick, 1997)) and other thinkers (see, for example, (Piaget, 1971)). In 1970 in his thesis written at Moscow State University, Burgin added a new dimension to this demonstrating that mathematics is a science of structures in the same sense as physics is a science of systems (see (Burgin, 1998)). Indeed, mathematics, as any natural science, has the theoretical, experimental and applied parts and all these three parts are dealing with structures.

In the theoretical part, mathematical structures, such as numbers, spaces or groups, are defined, mathematical statements are proved and theories of mathematical structures are developed. Algebra, geometry, calculus and topology are examples of such theories.

In the experimental part, mathematicians perform experiments with mathematical structures. For a long time, mathematicians used only their brains and paper for these experiments. Now computers are utilized more and more for finding regularities in the realm of mathematical structures.

In the applied part, mathematical structures are used to study various object, such as stars, atoms, planes, liquids, gases, organizations, animals, birds and many other things and systems. One researcher said that contemporary theoretical physics is a field of applied mathematics.

However, recent research shows that this is only an approximate understanding. More complete understanding tells us that mathematics is a science of systems of structures.

To grasp the difference, we have to find the difference between a (pure or abstract) structure and a system. In a pure structure, elements/parts do not have other properties and relations except those that belong to this structure (see Chapter 1 and Appendix). In a system, elements/parts may have other properties and relations. For instance, an abstract category is a (pure) structure, while the category of groups is a mathematical system (of structures) because its elements – groups - have inner structure. An abstract group is a (pure) structure, while the group of $n \times n$ orthogonal matrices or $n \times n$ unitary matrices is a mathematical system (of structures) because its elements – matrices - have inner structure.

It is possible to read more about mathematical structures in (Bourbaki, 1948; 1957; 1960; Shapiro, 1997; Resnick, 1997; 1999; Burgin, 2011). Besides, any mathematical book describes many mathematical structures.

Structural vision fertilizes the3 whole mathematics, For instance, there is also a special direction in mathematics called *the structural theory of probability*, which addresses the interpretation of probability, often debated in the scientific community (Rocchi, 2003).

Mathematics is intrinsically connected to logic. On the one hand, the main tool of formal mathematics is reasoning, which is formalized and studied by logic. On the other hand, logic on its higher level is formalized and developed as a mathematical discipline called mathematical logic (Russell, 1908; Church, 1956; Kleene, 1967). Thus, there are many structures in logic in general and in mathematical logic, in particular. Here are some examples of logical structures.

Logic has three basic levels:

1. The level of terms, concepts and names used in logic. This is the level of the *logic vocabulary*.
2. The level of expressions built from terms, concepts and names and used in logic, such as statements, propositions and predicates. This is the level of the *logic language*.
3. The level of logical inference and reasoning, for example, deduction. This is the level of *logic calculi* and *logical varieties*.

Each level has its own basic structures.

Looking at the first level, we see names, concepts and terms as its elements. The outstanding American logician Alonzo Church (1903-1995) explains importance of names for mathematical logic, contemplating the concept of name as one of the basic terms used in logic. At the beginning of his book (Church, 1956), Church clarifies the main concepts of logic. At first, he elucidates what *logic* is as a discipline. Then he explains the meaning and usage of the concept *name* is in logic. This gives the first basic logical structure – a named set considered in the previous chapter. Logic, as any other discipline, extensively uses names and consequently, it implicitly uses named sets.

As it is demonstrated in Section 6, concepts and terms also have developed structures and are represented by these structures.

Looking at the second level, we see statements, propositions and predicates as its elements. In the classical logic, a *proposition* is a sentence that is either true or false. It is possible to treat a proposition as a function, with words in it considered as variables, and the value of the function being either truth or falsity, a truth function. For example, the proposition "x is blue" is true if and only if the object denoted by x is blue.

For a long time, propositions have been studied as minimal elements (atoms) of logic that are either true or false. Recently, logicians started to study *structured propositions* (King, 2011). However, even before the famous mathematician and philosopher Bertrand Russell (1872-1970) ascribed definite structure to propositions, regarding concepts as constituents of propositions (Russell, 1903). That is why many current structured propositions theorists attribute the idea of structured propositions to Russell.

Statements and predicates also have developed structures and are represented by these structures (Church, 1956; Burgin, 2011).

Looking at the third level, we see inference, e.g., deduction, systems its elements. A deduction system consists of deduction rules. Each deduction rule and consequently, each step of deduction has the following form

$$A \rightarrow B \tag{2.1}$$

or

$$A \vdash B \tag{2.2}$$

Here A and B are some expressions or sets of expressions from the language of the corresponding logic. For instance, taking as A two propositions "We live in the USA" and "The USA are situated in the Western Hemisphere of the Earth," we can deduce as B the proposition "We live in the Western Hemisphere of the Earth."

Thus, a deduction rule is (has the structure of) the fundamental triad $(A, \rightarrow, B)$.

A deduction is a sequence of applications of deduction rules.

$$A \rightarrow B \rightarrow C \rightarrow D \rightarrow E \tag{2.3}$$

where A, B, C, D, and E are groups of logical statements. As a result, any deduction incorporates the structure of a deduction rule in a more elaborate structure of deduction, which is usually a sequence of named sets.

To conclude, we consider the role of mathematics in human cognition. As many know, physicists are trying to create a theory of everything. However, this is impossible because physics studies only physical, that is, material systems. So, in the best case, physicists will be able to create a theory of everything on the level of physics. However, for example, people are studied not only on this level but primarily on the biological, psychological and social levels. This shows that a physical theory of everything can describe the world only from its material side.

At the same time, we already know that everything has a structure and, as a rule, not a single one. Thus, the general theory of structures is a theory of everything as its studies everything. Mathematics, for example, is such a general theory of formal structures. That is why mathematics studies everything by formalized methods and techniques, being a theory of everything on its structural level.

Chapter 3

STRUCTURES IN SCIENCE

Structures are omnipresent and prevalent in all walks of life. Consequently, they play a vital role actually in all natural and social sciences, as well as in humanities, life sciences, medicine and engineering. That is why even in a big book, it is impossible to present all structures studied in science. Thus, we give only some examples choosing important structures from various fields.

In some scientific fields, such as chemistry, and areas of practical activity, such as engineering, structures play a leading role. For instance, the spatial structure of atoms, chemical elements, and molecules determines many properties of these chemical systems. In engineering, there is a special direction called *structural engineering*. However, to many other fields and areas, structures come mostly from mathematics.

Analyzing ordinary material objects, which people take themselves to be committed in ordinary life, Koslicki (2008) demonstrates that structures and objects are inseparable, arguing that every composite material object has a structure, and the structure is literally a part of the object. She bases her representation of system structures on mereology representing objects as systems that consist of parts and relations between these parts, while these relations form the structure of the object. In her structure-based approach, Koslicki explains that material objects are separated into classes with objects of the same kind/type. For each kind *K*, there is a list of types of ingredients and a structure, which being imposed on the ingredients produces objects of this kind.

However, in spite of many similarities between structures and forms, Koslicki distances herself from Plato and Aristotle, insisting that structures as formal components come with none of the theological and teleological cobwebs of ancient *forms*.

While formalized abstract structures are studied by pure mathematics, formal and informal concrete structures are studied by all sciences (natural, social and life sciences), applied mathematics and philosophy. For instance, physicists study physical structures. To represent physical structures in a rigorous way, they use mathematical structures. Mathematical structures help physicists to better understand physical reality, provide resources for innovative physical theorizing, and alleviate making new discoveries in physics. As an example, we can take the history of the positron discovery. We remind that a positron is one of the basic subatomic anti-particles, which are rare in our universe and opposite to subatomic particles.

In 1928, the outstanding physicist Paul Dirac (1902-1984) introduced what was later called the Dirac equation, unifying quantum mechanics and special relativity and describing the properties of the electron. However, when the energy of an electron was calculated from this equation, it gave two answers, one positive and the other negative. The positive answer evidently corresponded to the ordinary negatively charged electron. In contrast to other physicists, Dirac did not neglect the negative answer suggesting that it had to be a positively charged particle with the same mass as had the negatively charged electron.

Nobody believed him. Many physicists made a mock of him. However, in 1932, Carl Anderson (1905-1991) discovered a particle with the same parameters as were predicted by the theory developed by Dirac. Anderson called this new particle *positron* and received a Nobel Prize for this discovery. Dirac also received a Nobel Prize for this discovery and theoretical achievements.

Returning to structures, we see that there are three types of structures in science: structures of studied phenomena (or more exactly, of models of these phenomena), structures of scientific knowledge and structures of scientific cognition. Note that as cognition is an information process, structures of scientific cognition are specific information structures (Burgin, 2010).

3.1. Physics, Astronomy and Engineering

As argue some physicists, physics studies not physical systems as they are but structures of these systems called *physical structures*. Contemporary physics treats the physical world as a net of interacting components (systems), where there is no physical meaning to the state of an isolated object. A physical system (or, more precisely, its contingent state) is represented by the net of relations it entertains with the surrounding objects. As a result, the physical structure of the world is identified with such a global net of system relationships.

As North (2009) writes, physics is supposed to be telling us about the nature of the world, while physical theories are formulated in a mathematical language, using mathematical structures, and this implies that mathematics is somehow telling us about the physical make-up of the world.

To represent physical objects and processes, physicists use structures of three kinds: *visual images*, *descriptions in natural languages* and *mathematical structures*. Visual images, as well as descriptions in natural languages, are used mostly for informal explanations (usually to non-physicists) and for supporting researcher intuition, which is accustomed to work with images. To represent physical structures in a rigorous way, physicists employ mathematical theories. Now theoretical physics has become completely mathematical utilizing advanced fields in mathematics, such as group theory, topology, functional analysis and even category theory.

Dirac (1931) wrote that a powerful method of advancing physics was one that drew on the resources of pure mathematics by attempting "to perfect and generalize the mathematical formalism that forms the existing basis of theoretical physics and after each success in this direction, to try to interpret the new mathematical features in terms of physical entities..."

As North (2009) writes, we are used to talking about the *structure* posited by a given theory of physics. For instance, relativity in physics is a theory about the structure of the universe, which is considered in the space and time, or as physicists say, in the spacetime, making one mathematical structure the *spacetime* from two physical phenomena – space and time. Special relativity posits one spacetime structure, while different models of general relativity posit different spacetime structures. Understanding structure in this sense seems important for understanding what physics is telling us about the world. For instance, Carlip, *et al*, (2011) study the small scale structure of spacetime, demonstrating that vacuum fluctuations break space up into a large number of causally disconnected regions.

As modern theories of physics are formulated in mathematical terms, to understand *physical structures*, we must know *mathematical structures* that represent (model) physical phenomena. However, in this chapter, we do not consider mathematical structures of physics because they demand special and sometimes very advanced mathematical knowledge, while our goal is to make this book appropriate for a wide audience of readers who like science and are interested in the world where they live. Thus, we mostly describe structures that have some visual representation although it is necessary to understand that these representations, as well as descriptions in natural languages, are rather imprecise and give only a general impression of more exact mathematical structures. We present only informal, mostly visual, structures as examples.

Physicists study structures on various levels and in diverse scales – from subatomic particles, such as protons and neutrons, through the structure of the whole universe. For instance, looking in the sky, we see a hierarchy of structures: from stars and planets to galaxies and, on much larger scales still, galaxy clusters, and enormous voids between galaxies. Thus, a fundamental problem in physical cosmology is the cosmological generation of structure (Bojowald and Skirzewski, 2008). The current understanding of structure formation in the early universe is mainly built on a magnification of quantum fluctuations in an initial vacuum state during an early phase of accelerated universe expansion.

That is why we start with the quantum level, where the atomic structure impacts all properties of atoms. Let us consider this structure. Atoms are made up of three types of particles electrons, protons and neutrons. These particles have different properties. Electrons are tiny and very light particles that have a negative electrical charge. Protons and neutrons are much larger and heavier than electrons. Protons have a positive electrical charge, while neutrons have no electrical charge. Each atom is made up of a combination of these particles. Protons and neutrons form the nucleus of the atom, while electrons are "spinning" around this nucleus, forming a "cloud", which is arranged in shells. The electrons of an atom are bound to the nucleus by the electromagnetic force. This is an attractive model because allows natural visualization (see Figure 3.1). However, this image gives a very approximate representation of the atomic structure. A much better representation is obtained by such mathematical structures as differential equations.

Going to the opposite end of the physical scale and looking at the *global structure of the universe*, we see that it is hierarchical, beginning from the stellar level (level of stars) and coming to the scale of superclusters and filaments.

Stars are organized into galaxies. A *galaxy* is a massive, gravitationally bound system of stars. They also contain stellar remnants, an interstellar medium of gas and dust, and an important but poorly understood component called *dark matter*. Astronomers estimate that there are more than 100 billion (10^{11}) galaxies in the observable universe.

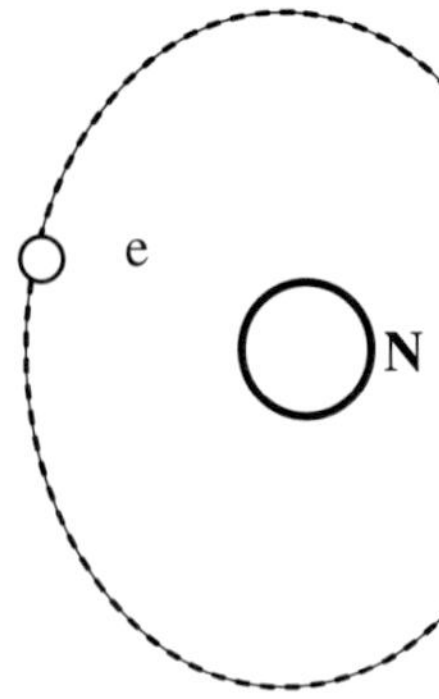

Figure 3.1. Image of an rough structure of a hydrogen atom where N is the nucleus and e is the electron.

Going from the structure of atoms to the structure of suatomic particles, physicists discovered that the proton, for example, is composed of three other fundamental particles: two up quarks (U) and one down quark (D) (see Figure 3.2).

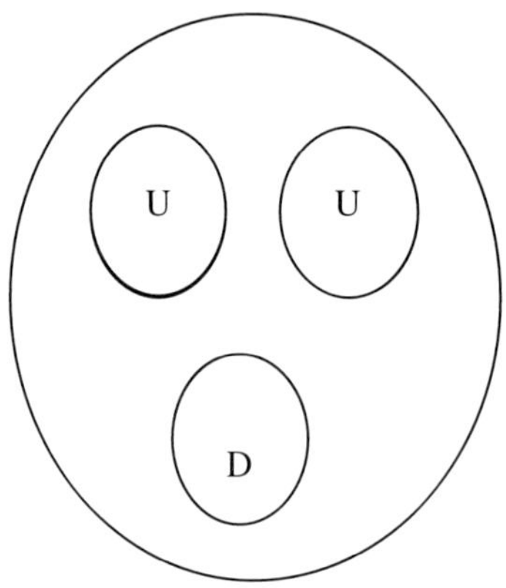

Figure 3.2. A rough structure of a proton.

The Sun with its planetary system is an element of the galaxy called the *Milky Way*, which contains about 200 billion (2×10^{11}) stars and has a form of a huge disc. The Greek philosopher Democritus (450–370 BC) was may be the first who suggested that the bright band on the night sky known as the Milky Way might consist of distant stars. Actual proof of this hypothesis came only in 1610 when Galileo Galilei used a telescope to study the Milky Way and found that it was composed of a huge number of faint stars.

Galaxies are often combined into *galaxy groups* and *clusters*, which are the largest known gravitationally bound objects arisen in the process of cosmic structure formation. galaxy groups are the smallest aggregates of galaxies, containing no more than 50 galaxies. Our own galaxy, the *Milky Way*, belongs to the Local Group of galaxies with more than 40 galaxies.

Clusters are larger than galaxy groups, containing up to several thousand galaxies. However, there is no sharp dividing line between the two. Clusters were formed relatively recently - around 10 billion years ago and continue to emerge now.

Superclusters are large groups of smaller galaxy groups, clusters and additional isolated galaxies, belonging to the largest known structures of the cosmos. Superclusters are so large that they are not gravitationally bound. Superclusters are elements of the largest structures of the universe, called *galaxy filaments*, *supercluster complexes*, *great walls* or *sheets*, which may span between several hundred million light-years to one billion light-years, covering more than 5% of the observable universe. Superclusters are separated by immense voids, creating a vast foam-like structure called the *cosmic web*.

Discoveries of superclusters and supercluster complexes started only recently in the late 1980s. Prior to 1989, it was commonly assumed that galaxy clusters were the largest structures in existence, and that they were distributed more or less uniformly throughout the universe in every direction. However, in 1989, Margaret Geller and John Huchra discovered the *Great Wall*, a sheet of galaxies more than 500 million light-years long and 200 million wide, but only 15 million light-years thick. The existence of this structure escaped notice for so long because it requires locating the position of galaxies in three dimensions, which involves combining location information about the galaxies with distance information from red shifts. In April 2003, another large-scale structure was discovered, the *Sloan Great Wall*. Larger than superclusters and filaments, there seems to be no continued structure, a phenomenon which has been referred to as the *End of Greatness*.

To understand the size of known clusters and superclusters, we remind that a light year is the distance covered by a beam of light in one year. It is a huge distance. For instance, the distance from the Earth to the Sun measured in light years is less than such a small number as 2 divided by a hundred thousand.

Some stars have planetary systems, which consist of planets and other relatively small celestial bodies that rotate around the star. For instance, now it is assumed that the Solar (planetary) system has 8 planets: *Mercury*, *Venus*, *Earth*, *Mars*, *Jupiter*, *Saturn*, *Uranus*, and *Neptune*. Before 2006, *Pluto* was also counted as a planet, but because Pluto is essentially less than any other planet and even less than the Moon, it was declassified and now it is called a *dwarf planet*.

Although any physical theory studies structures, there is also such a field as *structural physics*, which combines the branches of applied physics that study the physical phenomena and processes associated with the construction and use of various structures, such as buildings, bridges, towers, etc., developing appropriate structural design methods.

Structural physics includes such fields as:

- *construction heat engineering*, which studies and employs the laws ruling heat transfer in structures;
- *construction acoustics*, which studies the laws governing the transmission of sound in structures;
- *construction illumination engineering*, which studies and employs the laws governing the transmission of light in structures;
- the *theory of durability of structural members and building materials*,
- *structural climatology*,
- the *aerodynamics of structures*.

There is also a special branch of applied physics called *structural mechanics*, which deals with problems of the strength, rigidity, and stability of buildings and structures.

When we discuss physical structures, it is necessary to explain that now there is a tendency to change traditional physical structures in general and structures from quantum theory in particular to information structures. This approach comes from understanding that physicists study physical systems not directly but only through information they get from these systems. This resulted in creation of a school of thought about the role of information processing in physical processes and its influence on physical theories. According to one of the outstanding physicists of the 20th century John Archibald Wheeler (1911 – 2008), it means that every physical quantity derives its ultimate significance from information. He called this idea "*It* from *Bit*" where *It* stands for things, while *Bit* impersonates information as the most popular information unit (Wheeler, 1990; 1994). For Wheeler and his followers, space-time itself must be understood and described in terms of a more fundamental pregeometry without dimensions and classical causality. These features of the physical world only appear as emergent properties in the ideal modeling the physical reality based on information about complex interactions of very simple basic elements, such as subatomic particles.

This well correlates with one of the main neurobiological assumptions, which implies that things and the world perceived by people are constructed in their brains. This construction is based on the information received from these things and the whole world. Thus, whatever people know, in particular, all scientific knowledge, depends only on information they have. Mahoney (2010) asserts that science is not about nature, but about how scientists represent nature. According to him, we know about nature through the models scientists build of it, constructing them by abstraction from observation and experiments, manipulating them physically or conceptually, and testing their implications back against nature. The most developed models have the form of mathematical structures. However, any model (even a physical one) is some structure.

The most fundamental approach to physics based on information was developed by von Weizsäcker, his coauthors, and his followers in the, so-called, *ur*-theory (von Weizsäcker, 1958; Lyre, 1998). According to this theory, what physicists learn about nature comes from observation and experiments through measurement. At the same time, measurement is information extraction and so physical objects and processes are entirely characterized by the information acquired by measurement. In such a way, physics reduces to theoretical and experimental information acquisition and using this information for predicting measurement outcomes. Moreover, Resnikoff (1989) considers information as the subject matter for any kind of scientific investigation.

Data obtained from measurements give information that is used to produce (derive) knowledge about physical objects and processes. This implies that physical objects and processes are reduced to or even made of information. For instance, Born (1953) advocates that there are objects that exist independently of people but people can deal only with information about these objects and their structures. Based on these ideas, von Weizsäcker and his coauthors developed important parts of contemporary physics.

Like von Weizsäcker, Frieden (1998) developed a new approach to physics as a whole scientific discipline. It is based on the measure of information introduced by Ronald Aylmer Fisher (1890-1962), who was an English statistician, evolutionary biologist, and geneticist. Based on mathematical statistics, he introduced an original information measure now called Fisher information (Fisher, 1922).

The main assumption of this approach is that physics as a knowledge system about the universe is built from knowledge acquired through information reception and processing. Thus, it is possible to deduce laws of physics from information principles. According to this conception, the observer is included into the phenomenon of measurement as his properties as an information receiver (see, for example, (Herbert, 1987)) influence what data are collected and how they are collected and interpreted. Frieden (1998) demonstrates how, using regularities of information acquisition, to derive a great deal of contemporary physical theory.

This reduction of physical structures to information structures brought researchers to interesting results and novel ideas. For instance, Brukner and Zelinger (2002) were able to obtain resolution of many problems of quantum mechanics connected to the so-called measurement paradox when measurement of characteristics of one particle changes characteristics of another particle. However, when measurement is treated as information acquisition subject to laws of information, we come to the conclusion that it is not characteristics of another particle that are changing but our knowledge of these characteristics is changing. Our knowledge is reflected in the quantum state ascribed to the particle. Thus, acquisition of information about one particle gives us information about another, may be very distant particle, and this information changes the state of the second particle as our knowledge about that particle.

In addition, Muller (1998) suggested a formalization of physical structures. At first, he shows that physical theories are not identical to mathematical theories, and thus, physical structures are not identical to mathematical structures. After this, he builds an axiomatic system by extending Zermelo-Fraenkel axiomatic set theory with *urelementen*, also called *atoms* or *individuals*, where an *urelement* is an object (concrete or abstract) which is not a set, but that may be an element of a set.

Physics also has various structures of physical knowledge. On the top level, there are structures of physical theories and systems of such theories. They are studied in methodology of science and discussed in Section 3.9.

Such structures as classifications are very useful in physics. Here we give only one example of a physical classification, namely, a part of the modern classification of quantum objects. It is represented in Figure 3.3.

To conclude this section, it is necessary to mention such a field as *structural geology*, which studies the three-dimensional distribution of rock units based on their deformational histories. Structural geology utilizes measurements of present-day rock geometries to uncover information about the history of deformation (strain) in the rocks, and to understand the stress field that resulted in the observed strain and geometry of the rocks (van der Pluijm and Marshak, 2004).

Structures also play an important role in engineering. In particular, structures and *structural analysis* even form a separate area in engineering (see, for example, (Heyman, 1998; Martin, 1999)). Even more, there is a special part of engineering called *structural engineering*, which is a field of engineering dealing with the analysis and design of structures that support or resist loads (Heyman, 1999). Structural engineering is usually considered a specialty within civil engineering, but it can also be studied in its own right (WIKIPEDIA). There is a journal *Engineering Structures*, which encompasses such areas as: infrastructure engineering; earthquake engineering; structure-fluid-soil interaction; wind engineering; fire engineering; blast engineering; structural reliability/stability; life assessment/integrity; structural health monitoring; multi-hazard engineering; structural dynamics; optimization;

expert systems; experimental modeling; performance-based design; multiscale analysis; engineering economics. There is the Structural Engineering Institute in the United States and the Institution of Structural Engineers in Great Britain. For engineers, structures are not only objects with which engineers work but also tools in their work (see, for example (Burgin and Zak, 1982; 1984)).

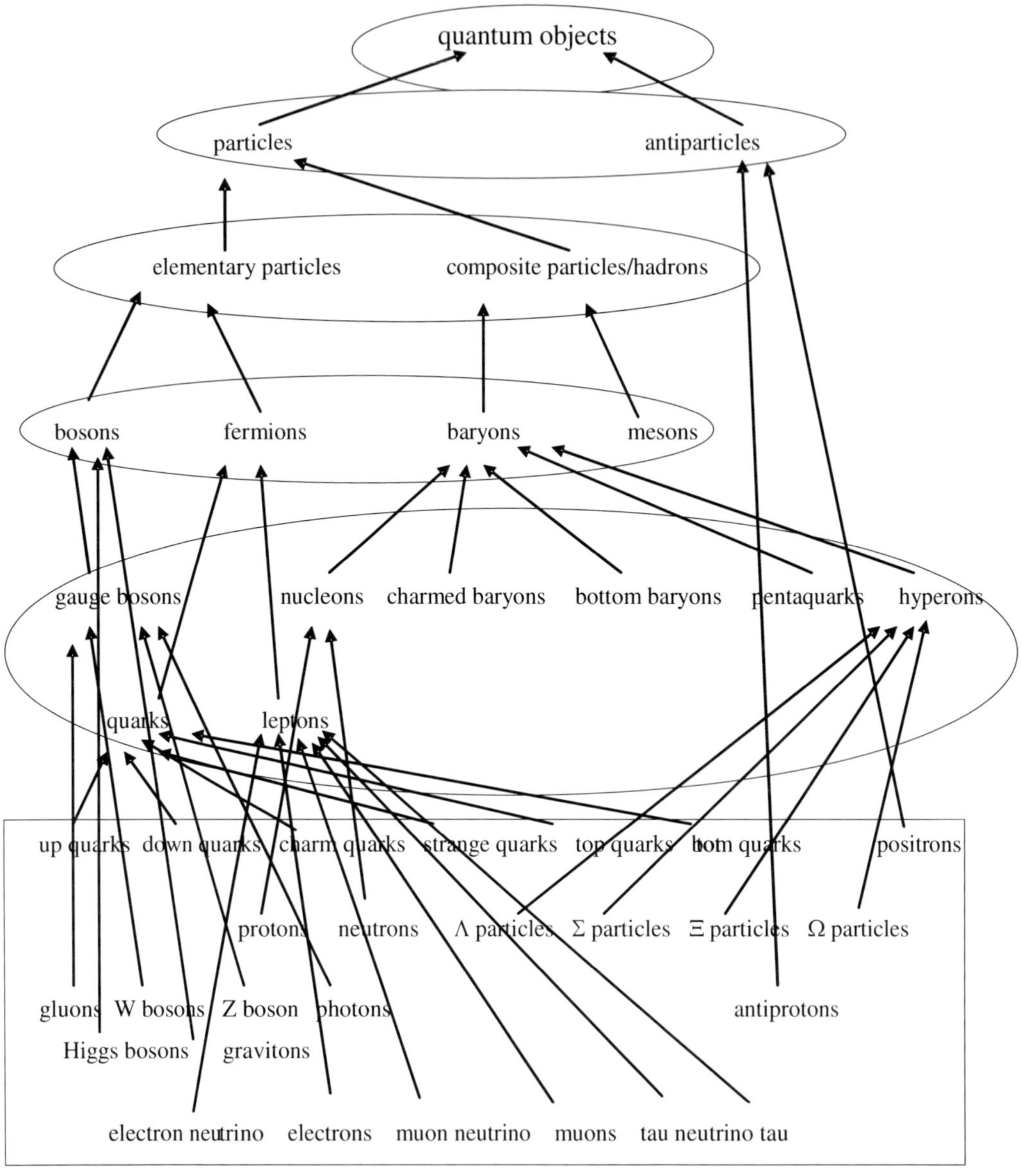

Figure 3.3. A part of modern classification of physical quantum objects (subatomic particles and antiparticles).

3.2. CHEMISTRY

As a branch of science, chemistry emerged from black magic and alchemy at the end of the 17th century. Usually alchemy was defined as the study of the composition of waters, movement, growth, embodying, disembodying, drawing the spirits from bodies and bonding the spirits within bodies.

From the end of the 17th century till the mid of the 19th century, chemistry existed as the traditional or classical chemistry. After this, it had become the modern chemistry.

In the 17th century, it was called *chymistry* and defined as the subject of the material principles of mixed bodies (Boyle, 1661). Later it acquired its contemporary name - chemistry and was defined as the art of resolving mixed, compound, or aggregate bodies into their principles; and of composing such bodies from those principles (Stahl, 1730) or as the science concerned with the laws and effects of molecular forces (Dumas, 1837). Now chemistry is defined as the science of substances: their structure, their properties, particularly with the properties of chemical bonds, and the reactions that change them into other substances

Structures became pivotal to chemistry from the very beginning. For instance, the concept of aggregates or units of bonded atoms, i.e. "molecules", traces its origins to Robert Boyle's 1661 hypothesis, in his famous treatise *The Sceptical Chymist*, that matter is composed of *clusters of particles* and that chemical change results from the rearrangement of the clusters. Boyle argued that matter's basic elements consisted of various sorts and sizes of particles, called "corpuscles", which were capable of arranging themselves into groups.

Later structures have become even more important. Now much of chemistry is explainable in terms of the structures of chemical compounds. The understanding of these structures depends very strongly on understanding of the electronic configurations of the elements.

A *chemical structure* includes *molecular geometry*, *electronic structure* and *crystal structure* of molecules. We remind that a *molecule* is an electrically neutral group of atoms held together by chemical bonds. The great scientist James Clerk Maxwell (1831-1879) defined a *molecule* as the smallest possible portion of a particular substance. Structures of molecules are studied and utilized both in physics and chemistry.

Electronic structure as a type of chemical structures describes the occupation of *molecular orbitals* of a molecule. Molecular orbitals are shells that electrons occupy in a molecule. This term reflects the idea that electrons revolve around a compact nucleus of the molecule.

A *crystal* is a solid material whose constituent atoms, molecules, or ions are arranged in an orderly, repeating pattern extending in all three spatial dimensions. The *crystal structure* as a type of chemical structures is a unique arrangement of atoms, molecules, or ions in a crystal. The crystal structure has an essential influence on the properties and behavior of the crystal

Molecular geometry as a type of chemical structures refers to the spatial arrangement of atoms in a molecule and the chemical bonds that hold the atoms together. Molecular geometry can range from the very simple, such as diatomic oxygen or nitrogen molecules, to the very complex, such as protein or DNA molecules. Molecular geometry can be roughly represented using a *structural formula* or *electron dot diagram*.

The *structural formula* of a chemical compound is a graphical representation of the molecular structure showing how the atoms are arranged.

An *electron dot diagram* displays the same structure as the corresponding structural formula but with more details.

In comparison with structural formulas and electron dot diagrams, a *molecular formula* gives more approximate structure of a molecule describing only the atoms that belong to the molecule, their number and indicating that all belong to the same molecule.

Here are some examples.

$$H - C \equiv C - H$$

Figure 3.4. The acetylene structural formula.

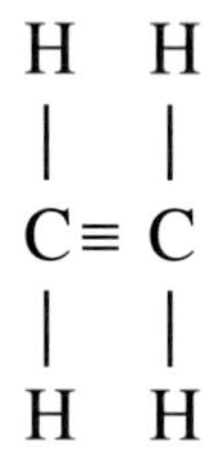

Figure 3.5. The ethylene structural formula.

Structures of chemical elements and their combinations, such as molecules, are very important in chemistry. Figure 3.6 gives some examples of used in chemistry chemical structures in symbolic forms. Here symbols are represented by images on the paper or images on the screen.

$$\text{a) } H_2O \qquad \text{b) } H-O-H \qquad \text{c) } H{:}\overset{\cdot\cdot}{\underset{\cdot\cdot}{O}}{:}H$$

Figure 3.6. Chemical structure of water.

(a) the structure of a molecule of water in the form of a chemical formula;
(b) the structural formula of water;
(c) the *electron dot diagram* of a molecule of water

Although structural formulas give you some idea of the structure of a molecule, they do not show the atoms' exact arrangement. For example, the methane molecule is actually a three-dimensional pyramid with four faces.

It is necessary to understand that structural formulas do not represent the three dimensional shapes or configurations of a molecule, which is also an important characteristic. Three dimensional configurations are best viewed with the aid of material models.

An important field in chemistry is *structural chemistry*, which includes the theory of molecular structure and chemical bond, the use of physical methods to examine the electronic and spatial structure of chemical species, structural features of liquids, solutions, surfaces and solid materials, and the crystal structure of solids. Structural chemistry develops new methods of synthesis, preparation of new compounds and materials, developing and implementing structural investigation methods by using X-ray, electron and neutron diffraction, scanning and transmission electron microscopy and various spectroscopic methods to the study of the physical, chemical and mechanical properties of the materials, exploring their applications, for example, in catalysis, energy storage and generation, molecule separation and as thermoelectric materials. It includes structural modelling and prediction from inorganic compounds over organic networks to proteins, as well as exploration of the relationship between structure and properties of chemical compounds.

Chemistry as a science also has various structures of chemical knowledge. On the top level, there are structures of chemical theories and systems of such theories. They are studied in methodology of science and discussed in Section 3.9. Chemical knowledge is often represented in the form of such structures as classifications. Here we give only one example of a chemical classification, namely, the top level of the classification of chemical elements. It is represented in Figure 3.7.

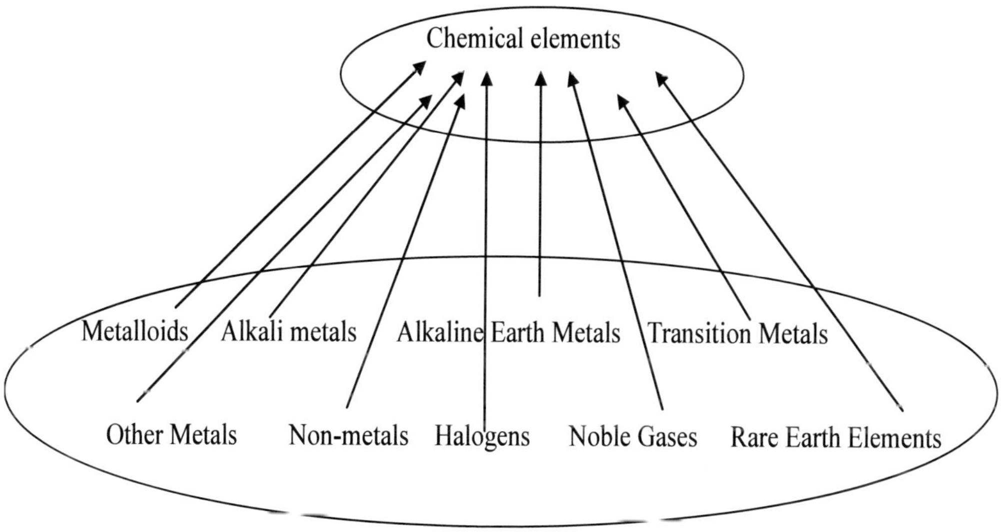

Figure 3.7. A classification of chemical elements in the form of a fundamental triad.

3.3. Biology and Medicine

Structures also play important roles in biology. For instance, one of the most famous discoveries of genetics was the discovery of the spatial structure of DNA in the form of a double helix. The whole DNA structure contains genetic information, which determines how parts of the body will be formed, where the parts will go, how growth and aging will proceed, and what influences from the past will be imprinted through the brain onto the mind. The entirety of an organism's hereditary information is called the *genome* (Sinden, 1994). It is stored in the linear sequence of nucleotides (see Figure 3.8) in DNA (deoxyribonucleic acid) and written as a text in the alphabet of three elements called *codons*, where a *codon* is a sequence of three nucleotides, or more exactly, of nucleotide bases, which together form a unit of genetic code in a DNA or RNA molecule.

Investigating the structure of DNA, researchers found that DNA in the human genome is arranged into 23 different types of *chromosomes*, or more exactly, 22 different types of chromosomes called *autosomes*, each present as two copies, and two *sex chromosomes*. Each chromosome contains many *genes*, the basic physical and functional units of heredity. Genes are specific sequences of DNA components called *bases*, which encode instructions on how to make proteins. There are four bases: *adenine* (A), *thymine* (T), *cytosine* (C), and *guanine* (G).

Structurally, DNA consists of two long polymer strands of simple units called nucleotides and their backbones (Sinden, 1994). Nucleotide bases are used to code information. This information is translated into proteins (amino acid sequences) by living cells.

Each nucleotide consists of three molecules and has the following structure.

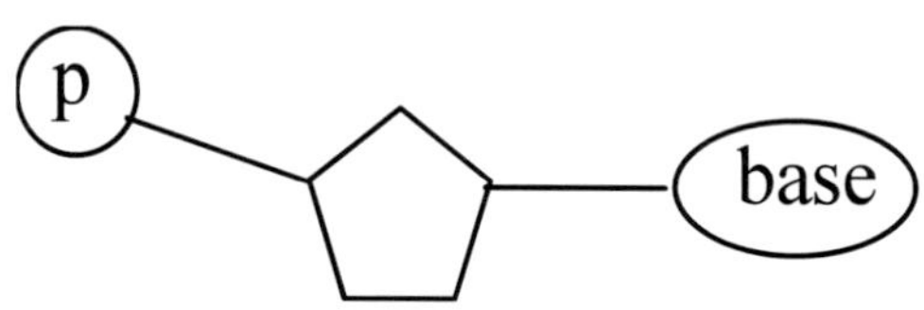

Figure 3.8. The nucleotide structure/

The structure of a part of the DNA strand is presented in Figure 3.9.

Base pairing form DNA is a structure that consists of two strands. In the canonical Watson-Crick base pairing, adenine (A) forms a base pair with thymine (T), as and guanine (G) with cytosine (C) in DNA. A part of DNA is presented in Figure 3.10.

As it is demonstrated in (Burgin, 2011), all these structures are fundamental triads (named sets).

In addition, functioning of living organisms is based on information transmission: information from DNA goes to RNA and from RNA it goes to proteins. This process is represented by the following named set chain, which is reproduced in many books on biology:

$$\text{DNA} \xrightarrow{\text{transciption}} \text{RNA} \xrightarrow{\text{translation}} \text{Protein}$$

Because RNA plays the role of intermediary, this named set chain may be treated as one named set.

$$\text{DNA} \xrightarrow{\text{RNA}} \text{Protein}$$

As one more example of biological structures, let us consider the structure of a cell.

The *cell* is the basic structural and functional unit of all known living organisms, which have a sophisticated modular structure. At the same time, a cell has a developed structure, blocks of are called organelles, each of which has a unique function, allowing the cell to function properly.

The structure of a cell comprises three basic components called a *nucleus*, *cytoplasm* and *surface*. A cell may have several nuclei, each of which has a spherical shape, includes a *membrane* and contains *chromosomes*, carriers of genetic information.

The cytoplasm of a cell has nine types of components, while the cell surface consists only of two components – *cell wall* and *plasma membrane*.

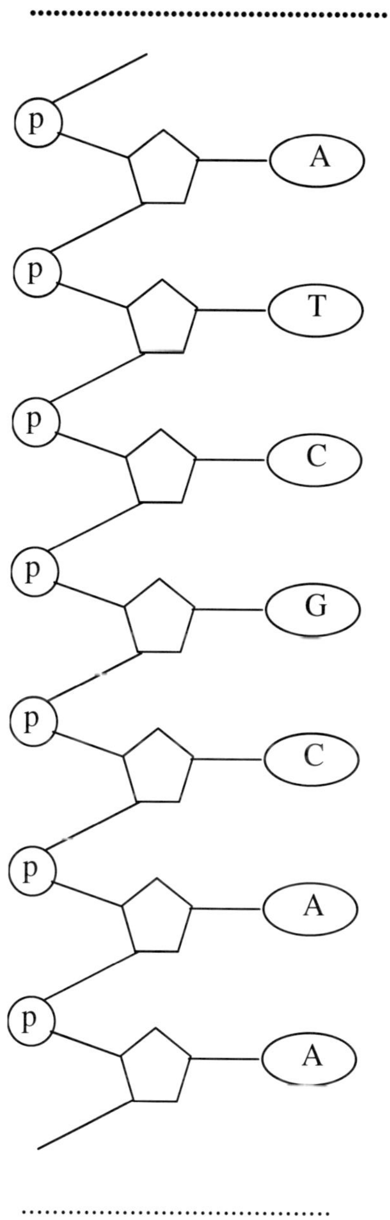

Figure 3.9. The DNA strand structure.

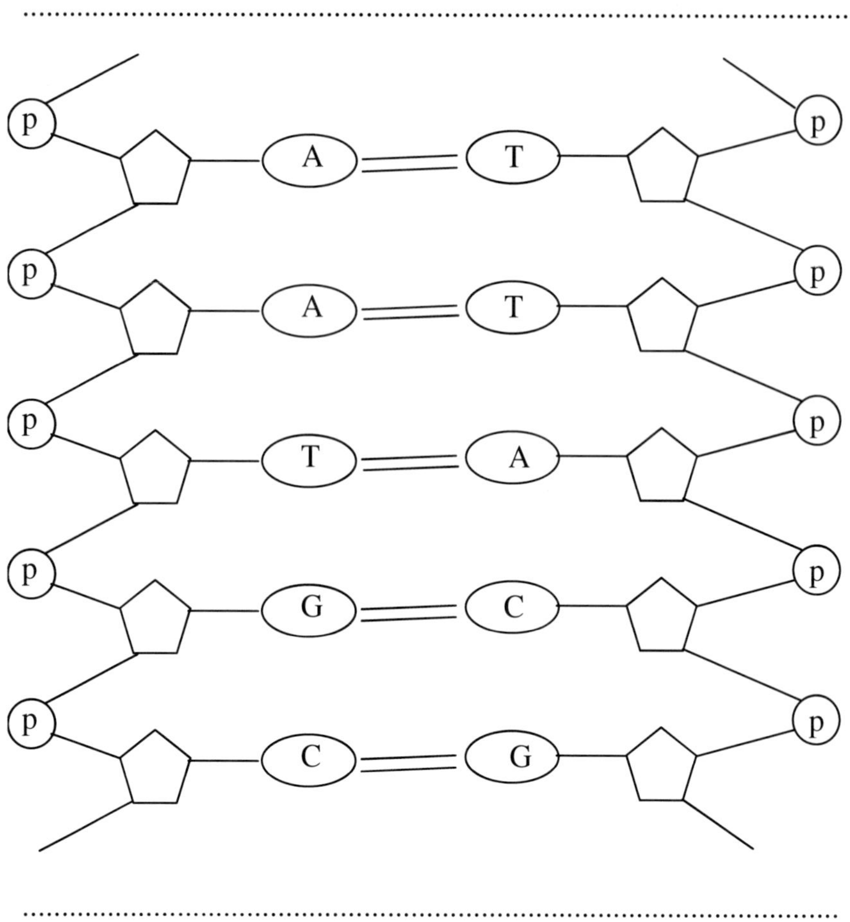

Figure 3.10. The DNA structure.

In biology, there is even a special branch called *structural biology*. It is a branch of molecular biology, biochemistry, and biophysics concerned with the molecular structure of biological macromolecules, especially proteins and nucleic acids, how they acquire the structures they have, and how alterations in their structures affect their function. This subject is of great interest to biologists because macromolecules carry out most of the functions of cells, and because it is only by coiling into specific three-dimensional shapes that they are able to perform these functions. This architecture, or the "tertiary structure" of molecules, depends in a complicated way on the molecules' basic composition, or "primary structures."

Biology also has various structures of biological knowledge. On the top level, there are structures of biological theories and systems of such theories. They are studied in methodology of science and discussed in Section 3.9.

Such structures as classifications are very useful in biology. Here we give only one example of a biological classification, namely, the classification associated with the genus *Homo*. It is the biologically determined cluster of living beings that include modern humans and species closely related to them (see Figure 3.11). This genus is estimated to be about 2.3 to 2.4 million years old, evolving from australopithecine ancestors with the appearance of *Homo habilis*. All species of the genus except *Homo sapiens* (modern humans) are extinct.

Homo neanderthalensis, traditionally considered the last surviving relative, died out 24,000 years ago, while a recent discovery suggests that another species, *Homo floresiensis*, discovered in 2003, may have lived as recently as 12,000 years ago.

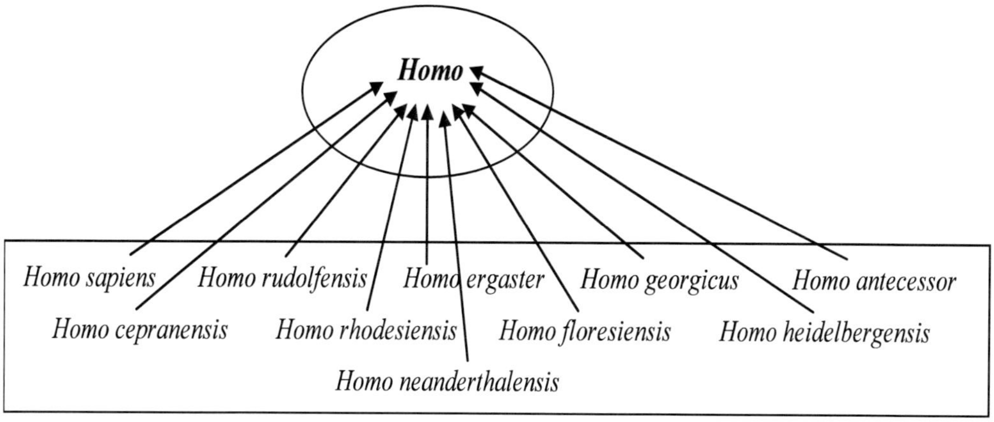

Figure 3.11. A part of the genus *Homo* structure.

One more field that is related to biology and explicitly utilizes the structural approach is *structural medicine*. It combines well-established and proven evaluation tools found in the mainstream medicine with the latest scientific developments in alternative treatment therapies, such as manual therapy.

3.4. Computer Science, Information Technology and Artificial Intelligence

Structures are extremely important in computer science, programming, and computer technology. Data structures form the core of information technology. It is impossible to build a database without explicit construction of data structures (see, for example, (Date, 2004; Tamassia, 1996; Burgin, 2008)). Data structures are extremely important for algorithms and software industry (see, for example, (Wirth, 1975; 1986; Burgin and Eggert, 2004)). To emphasize importance of data structures, Wirth (1975) wrote the book "Algorithms + Data Structures = Programs", while Michalewicz (1996) wrote the book "Genetic Algorithms + Data Structures = Evolution Programs." Computer architecture is a structure of a computer. Structural analysis is used in software technology (Burgin and Chelovsky, 1982; 1984). Researchers study knowledge structuration for the Web (see, for example, (Scharffe and Ding, 2006)). Structural representation of shapes is one of the most popular approaches in computer image recognition (Banerjee, *et al*, 1990). Structural characteristics programming languages influence their efficiency, flexibility and reliability (Burgin and Eggert, 2004).

In computer programming, there is a direction called *structured programming*. It is a programming technique that enforces a logical structure on the program being written to make it more efficient, manageable, understandable and reliable (Dahl, *et al*, 1972).

As an example, let us consider a *data structure*, which is a definite way of storing and organizing data in computers (Brass, 2008; Knuth, 1997). The goal of building or/and selecting an appropriate data structure is to efficiently organize information processing by designing effective algorithms. Different kinds of data structures are suited to different kinds of applications, and some are highly adapted to specific tasks. Many formal design methods and programming languages emphasize data structures, rather than algorithms, as the key organizing factor in software design.

The most popular data structures in computing are lists, records, arrays, and sets.

- A *list* or *sequence* is a data structure that implements a linearly ordered collection of symbolic objects called *values*, where the same value may occur more than once. An *instance* of a list is a computer representation of the mathematical concept of a finite sequence. Each *instance of a value* in the list is usually called an *item*, *entry*, or *element* of the list. If the same value occurs multiple times, each occurrence is considered a distinct item. Each element in the list has an *index*. The first element commonly has index 0 or 1 (or some other predefined integer). Subsequent elements have indices that are 1 higher than the previous element, providing an enumeration of the values in the list. There are *static* lists, which allow only inspection and enumeration of the values, and *mutable* or *dynamic* lists, which allow operation with items (values), e.g., it is possible to insert, replace or delete items (values) during the list's existence.
- An *array* stores a number of elements of the same type in a specific order, forming a vector (one-dimensional array), a matrix (two-dimensional array) or a higher dimensional matrix (n-dimensional array with $n > 2$). Elements from an array are accessed using a vector of integers, e.g., (3, 5, 8), to specify which element is required. For instance, if the array A has the form of (is) the matrix

$$\begin{pmatrix} 3 & 5 & 7 \\ 4 & 6 & 8 \end{pmatrix}$$

  then the vector (1,2) specifies the value 5 in the array A.
  There are *fixed-length arrays* or *expandable arrays*.
- A *record* (also called *tuple* or *struct*) is one of the simplest data structures. A record is a value that contains other values, typically in fixed number and sequence and typically indexed by names. The elements of records are usually called *fields* or *members*.
- A *set* is an abstract data structure that can store specific values, without any particular order, and no repeated values. Values themselves are not retrieved from sets, rather one tests a value for membership to obtain one of two values "in the set" or "not in the set".

Many other data structures are possible, but they frequently tend to be further variations and compounds of the above. Note that all these structures are named sets (Burgin, 2011).

The implementation of a data structure usually requires writing a set of procedures that create and manipulate instances of that structure. The efficiency of a data structure cannot be analyzed separately from those operations. This observation motivates the theoretical concept of an *abstract data type*, a data structure that is defined indirectly by the operations that may be performed on it, and the mathematical properties of those operations, often including their space and time cost characteristics.

For instance, implementation of the list data structure may provide some of the following operations:

- a constructor for creating an empty list, i.e., a list without elements
- an operation for testing whether or not a list is empty
- an operation for adding an entity to a list, e.g., prepending, i.e., adding an entity before the beginning of the list or appending, appending adding an entity to the end of the list
- an operation for determining the first component (or the "*head*") of a list
- an operation for referring to the list consisting of all the components of a list except for its first; this part is called the "*tail*" of the list.
- an operation for referring to first element, “head”, of the list

3.5. SCIENCE OF INFORMATION AND KNOWLEDGE

Understanding of the phenomenon of information is impossible without structures. In some models, information is presented as a kind of structures (Levich, 1978). Other models consider structures as domain of information action and existence (Burgin, 2010). In the context of the general theory of information, it is demonstrated that information is related to structures as energy is related to matter. This principle explains that information and structure are two basic phenomena of the world of structures (see Chapter 7).

It is interesting that defining information as “Information: that which determines form”, MacKay also related it to the World of Structures (MacKay, 1969) because *forms* form a special kind of *structures* (Burgin, 2010).

The unifying definition of information developed in (Burgin, 2010) is based on structures, i.e., information is inseparable from structures. First, information is structured in different ways. Second, any structure contains information in the proper sense. Third, information in the proper sense changes structures, e.g., cognitive structures such as knowledge.

In the 20th century, knowledge became the prime interest in several areas: in artificial intelligence, science, business, and technology to mention but a few. For instance, knowledge is now recognized by most businesses as a very important, valuable, and sometimes even critical resource. Good knowledge management helps companies to achieve their goals. Poor knowledge management results in mistakes, miscalculations and sometimes put company's future at risk.

The new strong emphasis on knowledge explicated importance of knowledge structure. Different methodologies have been developed to reconstruct knowledge structure and utilize it for different purposes, such as learning, knowledge management, knowledge representation and knowledge processing. For instance, knowledge is always represented by some structure and at the same time, knowledge always has its structure. Here we give two examples: the structure of an elementary knowledge unit (elementary “particle” of knowledge), which is developed in (Burgin, 2004a) and presented in Figure 3.12, and the structure of a top-level knowledge system (a knowledge "galaxy"), which is developed in (Burgin, 2011) based on the structure-nominative model (reconstruction) of scientific knowledge (Burgin and Kuznetsov, 1988; 1994; Balzer, *et al*, 1991) and presented in Figure 3.13.

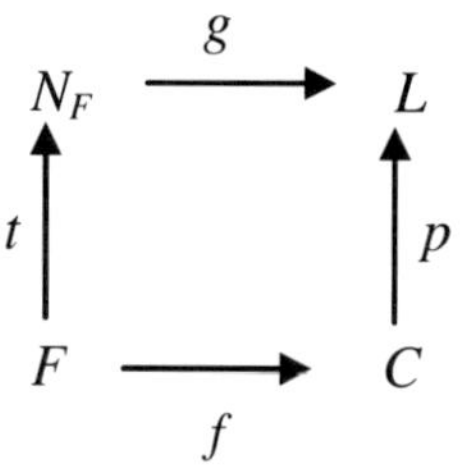

Figure 3.12. An elementary unit K of general knowledge about an object F.

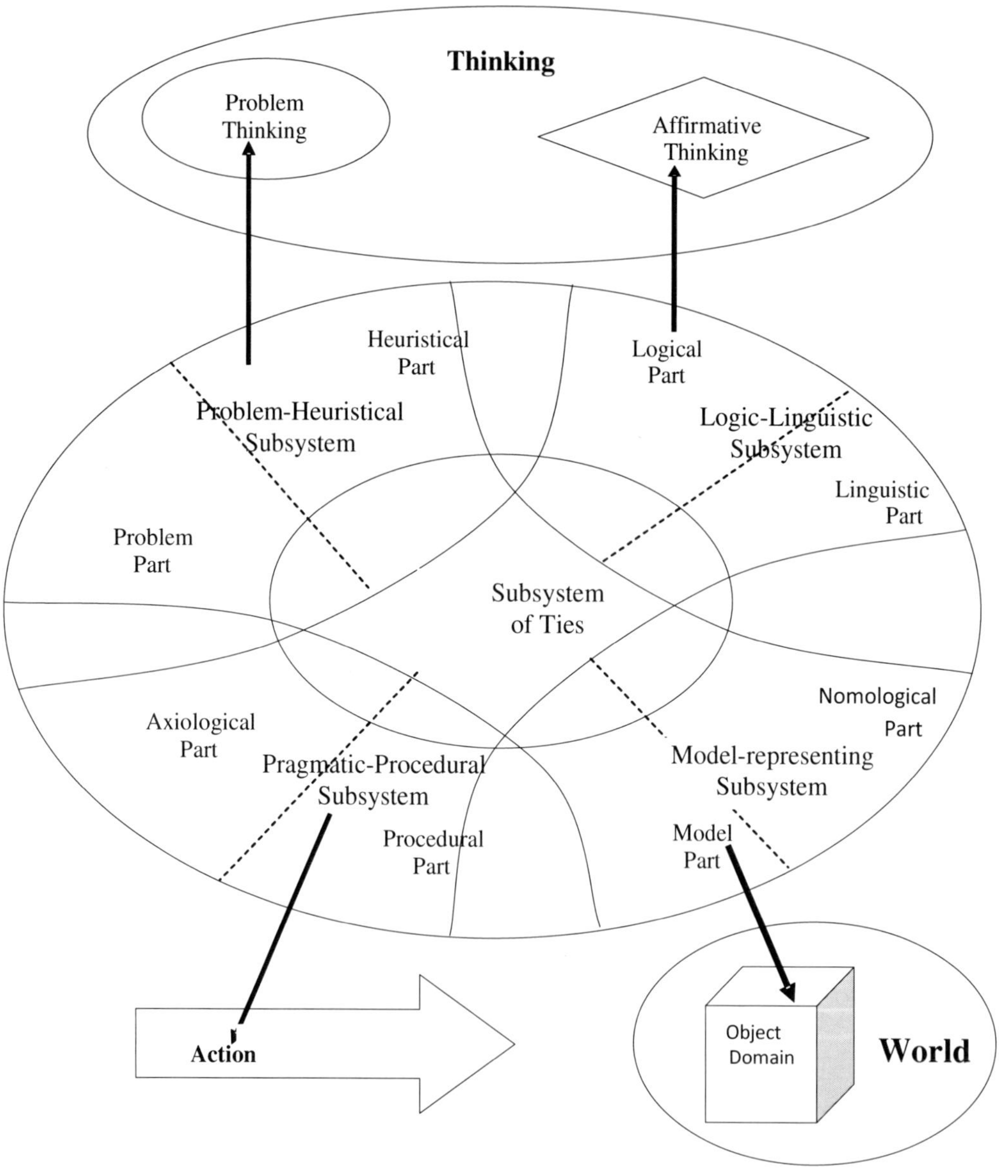

Figure 3.13. The structure of a advanced knowledge system with its external relations.

Here the correspondence t assigns the name N_F to the object F, the correspondence f assigns the property P (the intrinsic property) to the object F and the correspondence g assigns the value L (the ascribed property) of the property P to the name N_F. For instance, taking such property of material things as weight (the intrinsic property), in weighting any thing, we can get only an approximate value of the real weight, or weight with some precision (the ascribed property).

3.6. Anthropology and Sociology

Structural anthropology is based on Claude Levi-Strauss' idea that people think about the world in terms of binary oppositions, such as white and black, good and bad, high and low, big and small. Levi-Strauss assumed that every culture could be understood in terms of these opposites. "From the very start," he wrote, "the process of visual perception makes use of binary oppositions."

In such a way, Levi-Strauss, put structures in the center of anthropological studies, explaining that underlying structures - whether in kinship systems, myths, rituals or objects such as masks - determined the way the mind works. In particular, Lévi-Strauss examined kinship systems from a structural point of view, demonstrating how apparently different social organizations were in fact different permutations of a few basic kinship structures (1971). The leading idea was that the human mind operates in terms of such structures as binary oppositions and that such oppositions structure all the phenomena of human culture. Myths, for example, provide an imaginary resolution of the contradictions into which the binary ways of thinking lead people.

This type of thinking is related to classical logic, which has only two possible truth values – true and false. Now logics with multiple truth values, such as fuzzy logics or intuitionistic logics, are used, allowing better representation of knowledge and logical inference in many situations.

Structure is one of the basic concepts for sociology although it is not clearly conceptualized there (see, for example (Porpora, 1989)). Social structures explicitly underlie the majority of objects studied in sociology, such as: individuals and groups in definite relations to each other; relatively enduring patterns of behavior and relationship within society; social stratification; social institutions; social norms; and cognitive frameworks that determine actions of people, groups, and organizations. In this context, culture, religion, science, education, law, economy, and politics, all are social structures. Consequently, the notion of social structure is inherent for almost all central topics in social sciences. For instance, properties of the organization structure can determine organization's flexibility, capacity to change, potential for development and many other factors, and is therefore an important issue in management.

For social institutions and social groups, their structures are more important than their physical realization. Social relations are structures. Social forms in the sense of Georg Zimmel (1858-1918) are patterns of structures (Zimmel, 1964). In such a way, structures become explicit, and sociologists study them. There are sociological theories in which *structure* is the basic category and tool. For instance, Giddens (1984) developed a *structuration theory* of society, according to which social structures are the fundamental medium of human activities. *Structural functionalism* is a broad perspective in sociology and anthropology which interprets society as a structure with interrelated parts, concentrating on the positive and negative functions of social structures and aiming to explain the role of social structures and institutions in society, the relationship between these structures, and the manner in which these structures constrain the actions of individuals (Ritzer, 2007). According to structural functionalism, individuals have little to no control over the ways in which particular social structures operate. As a result, structural functionalists understand individuals in terms of their social positions.

Importance of structures for organizations is imprinted in the definition of organization. For instance, in (Daft, 1997) organization is defined as a social entity that is goal directed, designed as deliberately *structured* and coordinated activity system and is linked to the external environment. Note that coordination is a *dynamic structure*.

On the top structural level of society, Searl (2010) analyzes the structure of human civilizations through the exploration of the structure of social institutions. He describes this structure as generated by status functions and rules that define these functions. According to Searl, status functions are generated by speech acts. As a result, language assumes a central role in the construction of social structures.

Informally, the idea of a structure is generally understood in terms of relations that exist between the elements of some system, or more exactly, structures are built of relations (see Chapter 1 and Appendix). When we analyze relations, we study structures. Thus, the relational approach in modeling and study of the social reality explicates basic social structures in the form of relations between people. One of these directions developed a typological approach to social behavior because in their study of relations and interactions between people, sociologists came to the conclusions that it is possible to divide all relations into several patterns (models) in such a way that nearly all interactions, from close relationships to casual and distant ones, between people are organized by these patterns. This approach called *relational models theory* studies the four fundamental forms (patterns) of social relationships: *Communal Sharing*, *Authority Ranking*, *Equality Matching*, and *Market Pricing* (Fiske, 1991).

Extended relational models theory is a further development of the relational models theory. It describes the three pairs (dyads) of fundamental forms (patterns) of social relationships (Burgin, 2011). The first dyad consists of *Communal Sharing* (CS) and *Communal Differentiation* (CD). The second dyad consists of *Authority Ranking* (AR) and *Justice Evaluation* (JE). The third dyad consists of *Exchange Coordination* (EC) and *Exchange Pricing* (EP). Models, or patterns, that constitute each dyad are complementary. Communal Sharing emphasizes and is based on unity of people, while Communal Differentiation implies separation in some aspects and issues. Authority Ranking is based on subordination and subjugation, while Justice Evaluation tends to introduce some basic equity in society. Exchange Coordination reflects a huge diversity of resources and commodities, while Exchange Pricing reduces everything to money.

The relational approach allows one to explain the dynamics of the social development stages by the complexity of the six fundamental social relations. In the history of mankind, the simplest relationship pattern, Communal Sharing, was changed to a more complex Communal Differentiation. In turn, Communal Differentiation was changed to a simpler Authority Ranking, which offered higher stability in society. However, incongruence of particular features of Authority Ranking brought Justice Evaluation, which had even higher complexity, to the domineering position. To reduce this complexity, people started looking for justice only in exchange activities. This resulted in the key role of Exchange Coordination relationships. However, as the quantity of objects that are exchanged became extremely big and these objects had a diversity of types, people tried to elaborate a universal measure and equivalent of everything, making money the absolute measure - at first, of all things and then of (almost) everything. This transformation simplified relations between people, bringing society to Exchange Pricing. However, it is necessary to understand that this process was not linear going from one pattern of relations to the next higher pattern of relations. During some periods of time and in definite areas, transition of relation pattern can go backward, for example, Authority Ranking can change Justice Evaluation. Nevertheless, such reversals are temporary and the social process as a whole converges to higher and higher types of relation patterns.

Burgin (1998a) studied the structure of human culture, while different researchers studied the role of structures in art (Gryakalov, 1989; Pullan and Bhadeshia, 2000).

3.7. Economics

Such a direction as *structuralist economics* emphasizes the importance of structural features of studied phenomena. The approach originated with the work of the Economic Commission for Latin America and is primarily associated with its director Raul Prebisch and Brazilian economist Celso Furtado. Early structuralist models represented the productive structure of various countries and its interactions developing countries had with the developed world. The alleged declining terms of trade of the developing countries described by the Prebisch-Singer hypothesis, played a key role in this approach.

Dutt and Ros (2003) explain that structuralist economists try to identify specific characteristics of the structure of developing countries in order to assess the way economies adjust and their responsiveness to development policies. A general assumption in this approach is that the price mechanism fails

- as an equilibrating mechanism,
- to deliver steady growth,
- to produce a "desired" income distribution.

There is also the neostructuralist approach in economics which is based on the following principles (Bitar, 1988):

- importance of political and institutional factors in the analysis of economic problems.
- the need to raise the level of domestic saving in order to raise the rate of investment given that external sources of finance are likely to be hard to come by
- the false nature of dilemmas between, for example, planning and the market or agriculture and industry.
- the need to strengthen the productive and technological base.
- the importance of trying to improve the terms on which countries are integrated into the global economy and to improve international competitiveness.

- structural adjustment as only one component of structural change.

More recent contributions to structuralist economics have highlighted the importance of institutions and distribution across both productive sectors and social groups. A popular structural model of an industrializing economy has three commodity markets (food, manufactures and capital goods), foreign trade and income distribution which underpin the specification of a financial-sector with savings, investment, fiscal and monetary balances (FitzGerald, 1990).

In general, economics more and more uses mathematical models. As a result, mathematical structures acquire the leading role in economical theories.

Information structures also play an important role in information economics (Burgin, 2010). These structures represent the situation (state of the world) where customers/users are operating.

Business is vital for economy in modern society. Structural characteristics determine success in business. To give a scientific explanation of these relations, Buckley (1990) studied the structure of business.

3.8. Psychology

Structures are also explicitly used in various psychological theories. A *psychological structure* refers to a relatively stable organization of component parts that perform one or more related psychological functions. Let us consider some psychological structures.

Psychoanalysis is one of the basic psychological directions. It was founded and developed by Sigmund Freud (1856-1939), his collaborators and students at the end of the 19th and the first part of the 20th centuries. Freud's former students, such as Alfred Adler (1870-1937), Carl Gustav Jung (1875-1961), and Wilhelm Reich (1897-1957), contributed to the further extension and growth of psychoanalysis, as well as of new directions rooted in it, such as Individual Psychology of Adler or Analytic Psychology of Jung. Later psychoanalytic theory was expanded, been criticized and developed in different directions, by neo-Freudian psychologists, such as Erik Erikson (1902-1994), Erich Fromm (1900-1980), Anna Freud (1895-1982), Karen Horney (1885-1952), Harry Stack Sullivan (1892–1949) and Jacques-Marie-Émile Lacan (1901–1981). Now there is a diversity of theoretical orientations and approaches under the broad umbrella of psychoanalysis do.

One of the cornerstones of psychoanalysis is the structured model of the psyche brought forth by Freud in 1923 (see (Freud, 1949)). It consists of three parts - *Id*, *Ego*, and *Super-Ego* - and is presented in Figure 3.14 in the graphical form.

In this system, the *Id* is the irrational and emotional part, the *Ego* is the rational part and the *Superego* is the moral part.

According to Freud, the *Id* is an important part of personality, which is responsible for all needs and urges of a person and is present at birth as the repository of basic instincts, which Freud called *Triebe* (*drives*). It is unorganised and unconscious, operating mostly on the *Pleasure Principle* lacking realism or foresight. In this context, the *Id* wants feel good all the time, with no consideration for the reality of the situation or the needs of anyone else.

Contrary to the *Id*, the *Superego* is the ethical part of personality/psyche being responsible for ideals and moral. It develops due to the moral and ethical restraints placed on us by other people and employs self-observation, self-criticism and other reflective and judgemental faculties. Many equate the *Superego* with the conscience as it dictates our belief of right and wrong.

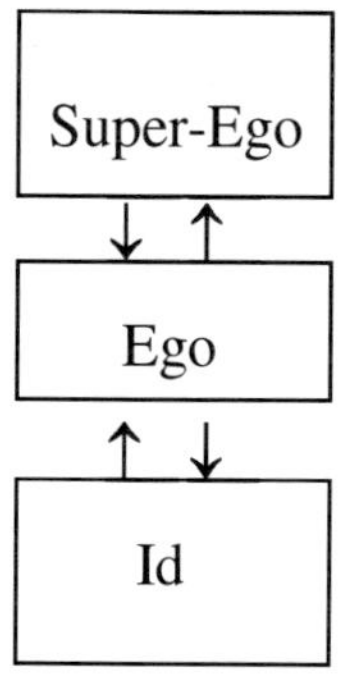

Figure 3.14. Freud's structural model of the psyche/personality.

The *Ego* is based on the reality principle, connecting the *Id* and the *Superego* and moderating between the demands of the *Id*, the *Superego*, and the reality of the external world. The *Ego* understands that other people have needs and desires and that sometimes being impulsive or selfish can hurt the person in the long run. The *Ego* is developing slowly and gradually, building its operation on the *Reality Principle*.

This model of personality develops a line of reasoning as a groundwork for explaining various (or perhaps all) psychological conditions, pathological and non-pathological alike. These conditions result from powerful internal tensions—for example: 1) between the *Ego* and the *Id*, 2) between the *Ego* and the *Superego*, and 3) between the love-instinct and the death-instinct. The *Ego* is caught between the *Id* and the *Superego*, finding itself simultaneously engaged in the conflict being pressed by the demands of the *Id* and the superior position of the *Superego*. At the same time, the interplay between the love instinct and the death instinct can manifest itself on any level of the psyche.

This shows that Freud's model of the psyche has the structure of a fundamental triad considered in Chapter 1. Indeed, in Freud's model, the *Ego* connects the *Id* with the *Superego*. (see Figure 3.15). The *Id* is also connects the personality to the reality, forming another triad.

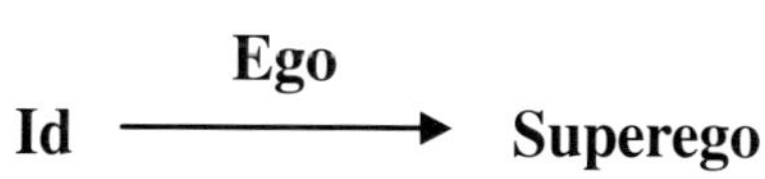

Figure 3.15. The linear form of Freud's structural model of psyche/personality.

In a healthy person, according to Freud, the *Ego* is the strongest so that it can satisfy the needs of the *Id*, not upset the superego, and still take into consideration the reality of every situation. Contrary to this, when the *Id* gets too strong, impulses and self-gratification take over the person's life. If the *Superego* becomes to strong, the person would be driven by rigid morals, would be judgmental and unbending in his or her interactions with the world.

In addition, Freud's model is built upon the presupposed existence of conscious and unconscious mentality as Freud (1949) states, "[About consciousness and the unconscious] there is nothing new to be said... the division of mental life into what is conscious and what is unconscious is the fundamental premise on which psycho-analysis is based." He further distinguishes between two types of unconscious thoughts: "preconscious" ideas, which are latent yet fully capable of becoming conscious; and "unconscious" ideas, which are repressed and cannot become conscious without the help of psychoanalysis. This gives an additional psychoanalytic model of the psyche presented in Figure 3.16 in the graphical form.

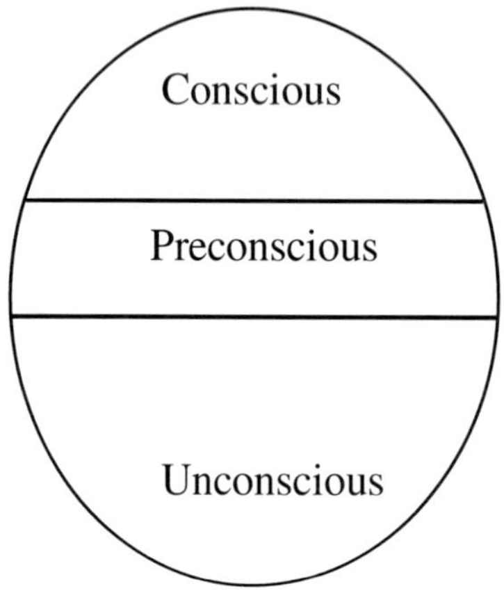

Figure 3.16. Additional Freud's model of the psyche.

It would be too simplistic to assume that the Unconscious and the Conscious map directly onto the *Id* and the *Ego*, respectively. In contrast to this, as Freud discovered, the supposedly conscious *Ego* possessed unconscious thoughts, for example, when *Ego* unknowingly resists to its own inclinations. Besides, the *Superego* also mostly belongs to the Unconscious.

The additional Freud's model of the psyche also has the structure of a fundamental triad because the Preconscious connects the Unconscious with the Conscious.

It is possible to read more about the structural traits of psychoanalysis in (Burgin, 1994b).

One more important psychological directions is *transactional analysis*, which is an integrative approach including elements of psychoanalytic, humanist and cognitive approaches. It was developed by the psychiatrist, Eric Berne (1910 –1970), during the late 1950s (Berne, 1964) and also uses many structures.

At any given time, a person is in an ego-state, experiencing and expressing itself through a mixture of behaviours, thoughts and feelings. According to Berne, there are three ego-states (see Figure 3.17):

- *Parent* (*exteropsyche*) is an ego-state when people behave, feel, and think as usually parents do. Often this is an unconscious mimicking of how their parents (or other parental figures) acted, or how people interprete their parent's actions. For instance, a person may shout at someone out of frustration because they learned such a behavior from an influential figure in childhood.

- *Adult* (*neopsyche*) is an ego-state resembling a computer processing of information with an objective appraisal of reality and without explicit emotions that could affect its operation. The goal of psychotherapy based on transactional analysis is to teach the patient how to strengthen the Adult.
- *Child* (*archaeopsyche*): a state in which people behave, feel and think as children do. For instance, a person who receives a poor evaluation at work may respond by looking at the floor, and crying or pouting, as they used to when scolded as a child. Conversely, a person who receives a good evaluation may respond with a broad smile and a joyful gesture of thanks. The Child is the source of emotions, creation, recreation, spontaneity and intimacy.

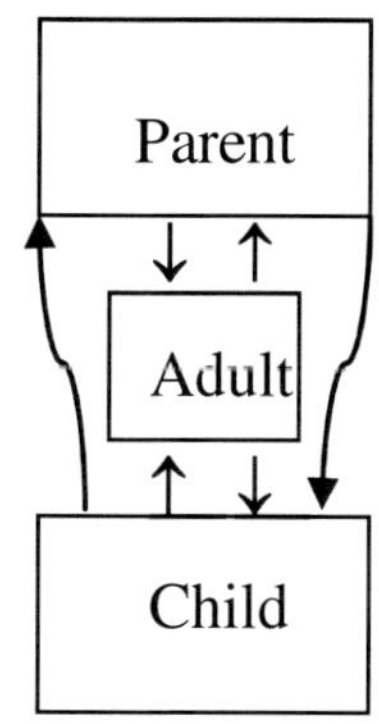

Figure 3.17. Berne's structural model of personality.

The Berne's model of the personality has the structure of a fundamental triad considered in Chapter 1 because *Adult* connects *Child* with *Parent*.

Berne differentiated his *Parent*, *Adult*, and *Child* ego-states from actual adults, parents, and children, by using capital letters when describing them. These ego-states may or may not represent the real relationships between people. For instance, in the workplace, an adult supervisor may take on the *Parent* role, and scold an adult employee as though she was a *Child*. At the same time, a child, using her *Parent* ego-state, could scold her actual parents as though the parents were a *Child*.

Within each of these ego states, there are additional structures.

Parent can be either *nurturing* (permission-giving, security-giving) or *criticizing* (comparing to family traditions and ideals generally in negative ways) or *distant* (Berne, 1964).

Child can be either more *natural* (*free*) or *adapted* to others, where the state *free* can be *positive* (I do what I want to do) or *negative* (I am not going to do what you tell me) (Berne, 1964; Burgin and Neishtadt, 1993).

Adult can be *autocratic*, *empathic* or *rational*, while rationality can be based either on *reasoning* or on a system of *rules* and *regulations* (Burgin and Neishtadt, 1993).

These structures categorize individual patterns of behavior, feelings, and ways of thinking, that can be *functional* (*beneficial* or *positive*) or *dysfunctional* (*counterproductive* or *negative*). Note that we have here a system of oppositions so extensively employed in structuralism (see Chapter 4 and Section 3.6)

Berne describes four types of diagnosis of ego states: the *behavioral diagnosis*, *social diagnosis*, *historical diagnosis* and the *phenomenological diagnosis* of ego states. For a complete diagnosis of an individual, one needs to complete all four types.

Ego states do not correspond directly to Sigmund Freud's *Ego*, *Superego* and *Id*, although there are obvious parallels, namely, the *Superego* corresponds to the *Parent*, the *Ego* corresponds to the *Adult*, and the *Id* corresponds to the *Child*. Ego states are consistent for each person and the particular ego state that a given person is communicating from is determinable by external observation and experience.

There are two basic processes/actions in the structures of transactional analysis.

- *Transactions* are components of the flow of communication, which occur simultaneously on both explicit and inner psychological levels.
- *Strokes* are the recognition, attention or responsiveness that one person gives another.

Strokes can be positive or negative ("cold prickles"). Our experience shows that people hunger for recognition, and that lacking positive strokes, they will seek whatever kind they can, even if it is recognition of a negative kind. People often create pressure in or experience pressure from others to communicate in a way that matches their style. For instance, a boss who talks to his staff as a controlling parent will often get back self-abasement or other childlike responses. Those employees who resist may get removed or labeled as "trouble".

Transactions also can be experienced as positive or negative depending on the nature of the strokes within them. However, a negative transaction is preferred to no transaction at all, because of a fundamental hunger for strokes.

The nature of transactions is important to understanding communication. There are basically three kinds of transactions:

1. *Complementary* or *symmetric transactions* preserve the same relation between participants in their interaction. For instance, when the first participant interacts with the second participant as the Parent to the Child, then the second participant interacts with the first participant as the Child with the Parent.
2. *Crossed* or *asymmetric transactions* do not preserve the same relation between participants in their interaction. For instance, it is possible the first participant interacts with the second participant as the Parent to the Child, while the second participant interacts with the first participant as the Adult with the Child.
3. *Duplex transactions* imply two different relations at the same time. For instance, the question "Where did you hide the scissors?" implies the relation Adult-Adult on the surface. However, the word *hide* brings forth the attitude of the Parent to the Child of the receiver.

Transactional analysis also considers six ways of structuring time by giving and receiving strokes: withdrawal, ritual, pastimes, activity, games and intimacy.

Withdrawal means no strokes are being exchanged

A *ritual* is a series of transactions that are complementary (reciprocal), stereotyped and based on social programming. For instance, two people may have a daily two stroke ritual, when each one greets the other with a "Hi" or “Hello”.

A *pastime* is a series of transactions that is complementary (reciprocal), semi-ritualistic, and is mainly intended as a time-structuring activity. *Pastimes* have no covert purpose and can usually be carried out only between people on the same wavelength. They are usually shallow and harmless. *Pastimes* are a type of smalltalk.

In the context of transactional analysis, *activities* mean the individuals work together for a common goal, such as work, sports or something similar. In contrast to *pastimes*, there is a meaningful purpose guiding the interactions, while *pastimes* are just about exchanging strokes. Strokes can then be given in the context of the cooperation. Thus, the strokes are generally not personal, but related to the *activity*.

A *game* is a series of transactions that is complementary (reciprocal), concealed, and proceeds towards a predictable outcome. *Games* are often characterized by a *switch* in roles of players towards the end. Berne identified dozens of *games*, noting that, regardless of when, where or by whom they were played, each *game* tended towards very similar structures in the number and roles of players, the rules of the game, and the game's goals.

Intimacy, as a way of structuring time, allows one to exchange the strongest strokes without playing a game. Intimacy differs from games as there is no covert purpose, and differs from *activities* as there is no other process going on which defines a context of cooperation. Strokes are personal, relating to the other person, and often unconditional.

Here we can see two triads. The first one consists of *withdrawal*, *ritual* and *pastime* where all transactions have no covert purpose. In the second one, which consists of *intimacy*, *activity* and *game*, transactions have some purpose.

Abraham Maslow (1908-1970) built his personality model in the form of a Hierarchy of Needs (Maslow, 1954). According to his theory, only when the lower order needs of physical and emotional well-being are satisfied, it becomes reasonable to go to the higher order needs of influence and personal development.

Maslow's original Hierarchy of Needs model was developed in 1940-50s and first published in (Maslow, 1943). At this time, the Hierarchy of Needs model comprised five layers of needs (Figure 3.18).

The *physiological and biological needs* include the needs people have for oxygen, water, protein, salt, sugar, calcium, and other minerals and vitamins, as well as the need to maintain temperature (98.6° F or near to it), to be active, to rest, to sleep, to avoid pain, and to have sex.

When the physiological needs are mostly satisfied, people come to the second layer developing *needs for safety and security*, which include stability, protection, order, organization, some boundaries.

When physiological needs and safety needs are mostly satisfied, people come to the third layer of *love and belonging needs*, which include the need for friends, a sweetheart, children, affectionate relationships in general, and even a sense of community.

After satisfying these needs, people come to *esteem needs*, which exist on two levels. The lower level includes the *need for the respect of others*, the *need for status*, *fame*, *glory*, *recognition*, *attention*, *reputation*, *appreciation*, *dignity*, even *dominance*. The higher level involves the *need for self-respect*, including such feelings as confidence, competence, achievement, mastery, independence, and freedom.

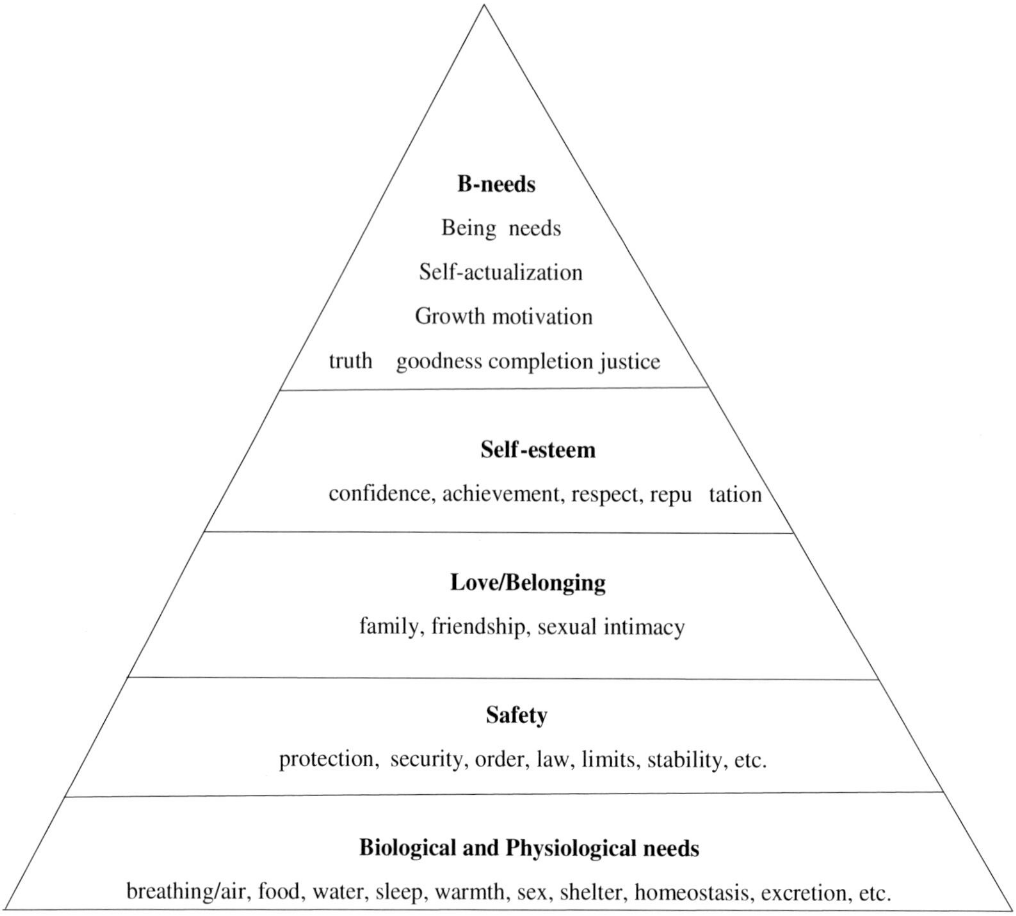

Figure 3.18. The original Maslow's Hierarchy of needs.

The bottom four levels in the Maslow hierarchy are *deficit needs*, or *D-needs*. In contrast to this, Maslow used a variety of terms to refer to the top level, calling it *growth motivation* (in contrast to deficit motivation), *being needs* (or *B-needs*, in contrast to D-needs), and *self-actualization*. Examples of B-needs are *truth*, rather than dishonesty; *goodness* in contrast to evil; *completion* in contrast to incompleteness; and *justice* in contrast to injustice and lawlessness.

Later Maslov added new classes of needs to his model without explicitly including them into the hierarchy, allowing other researcher to build two new extended hierarchies of needs represented in Figures 3.19 and 3.20.

The *extensive model* of personality developed by Burgin (1998b; 2010) also has the hierarchical structure and incorporates other systemic models of personality, both psychological and sociological.

According to the extensive model, the material component of personality consists of ten layers (levels), which are consecutively formed over the vital core of a person comprising psyche, soul, and spirit. According to the extensive model, personality has the structure represented in Figure 3.21.

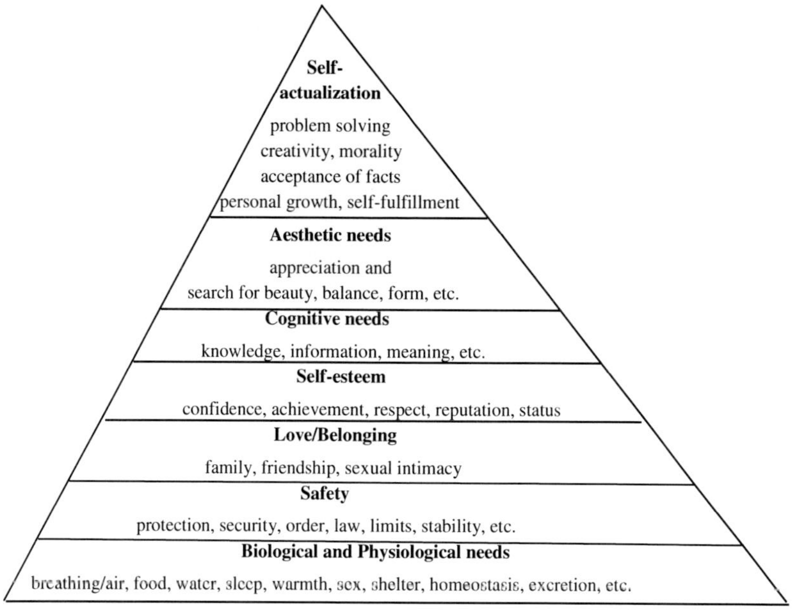

Figure 3.19. 1970s extended Maslow's Hierarchy of Needs.

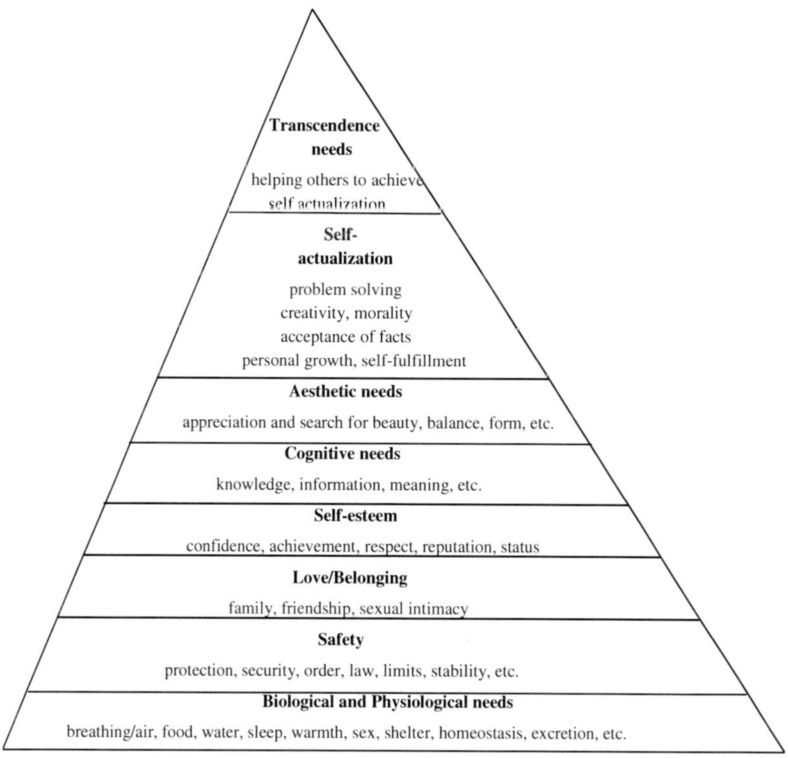

Figure 3.20. 1990s extended Maslow's Hierarchy of Needs.

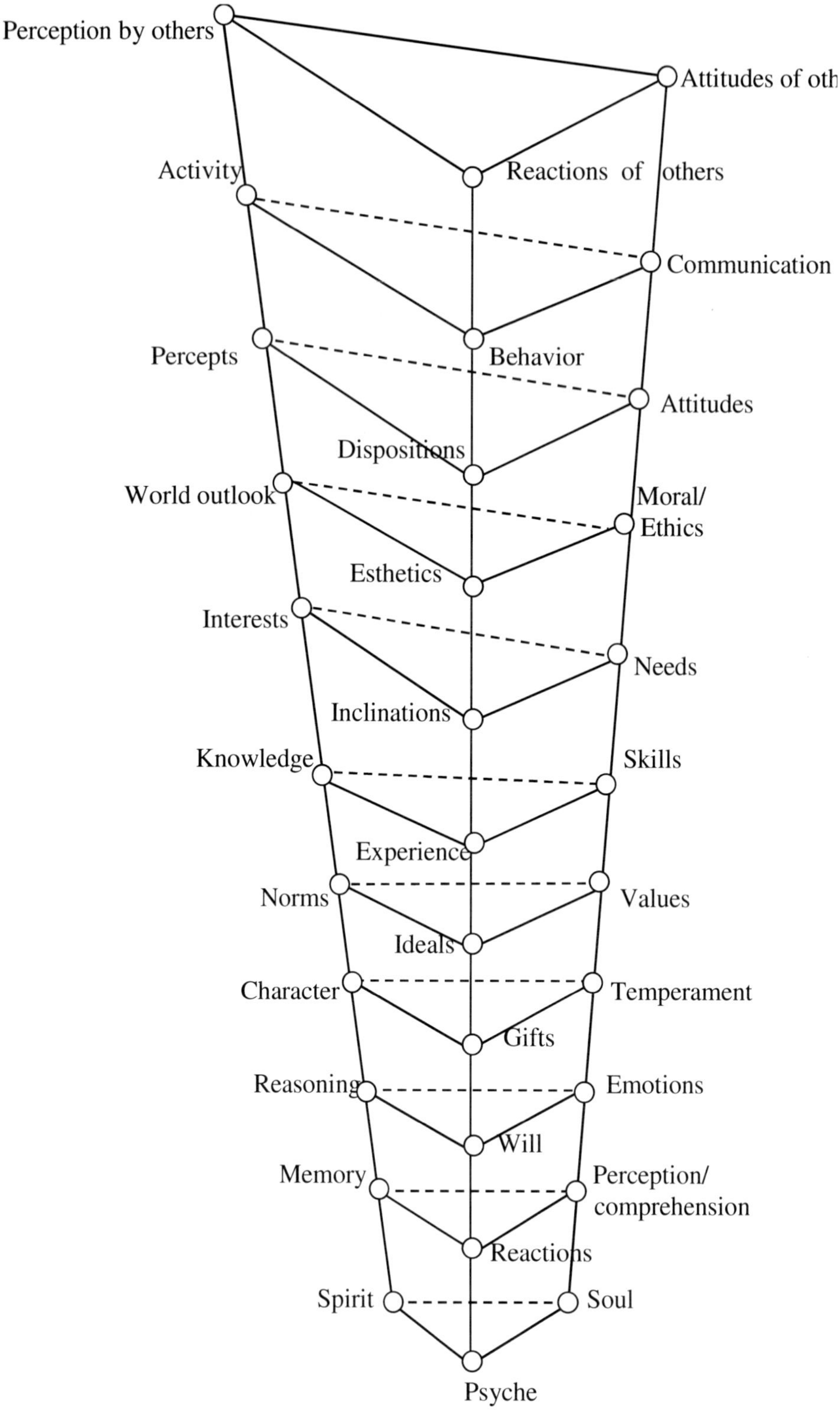

Figure 3.21. The Extensional model of personality.

Each layer of the model consists of three parts. Each part constitutes a definite system and represents some essential feature of a person. On the psychological level of personality, each part has three components: a system called the *structural brain scheme* (actual or virtual), a *process* that is realized in this system and through which the corresponding *feature* emerges, and what is produced by this system. For instance, intelligence of a person is a complex feature (trait) distributed in psyche and generated by collective functioning of different schemes in the nervous system of a human being. However, one of the schemes contributing to intelligence plays the leading role, while other systems have only supportive functions. We call this leading scheme the *intellect* or the system of *rational intelligence* of the individual.

Structures also play an important role in neuropsychology. For instance, Wilimzig, et al, (2012) demonstrate that the structure of intracortical inhibition contributes to the speed of human perceptual decision-making processes.

Warner (2000) studied the structure of human imagination.

3.9. Methodology of Science

The *structural approach* in methodology of science is very popular. At the beginning of the 19th century, Duhem published a book on the structure of physical theories (Duhem, 1906/1954). Later other authors studied structures of scientific theories from different perspectives. Suppe (1974) studied logical structures of scientific theories. Thompson (1989) studied the structure of biological theories. Kuhn (1962) studied the structure of scientific revolutions. Tondl (1998) studied the thematic structure of science.

Scheffler (1963) introduced the concept of a *structural term* to name terms that occur in the discourse about any science. Examples of such terms are *theory*, *law*, *result*, *discovery*, *hypothesis*, *observation*, *explanation*, *evidence*, and *proof*. Structural terms are used to form and to describe the structure of a scientific discourse.

Although many researchers, philosophers and scientists, studied structural peculiarities of scientific knowledge, a specific *structuralist direction* was initiated by Sneed (1979). From his perspective, scientific knowledge in general is a huge and complex conceptual network of scientific theories. In this context, scientific knowledge exists on different levels of granularity and a scientific theory is a collection of models in a form of various mathematical structures, for which the set-theoretical language is used. The basic level contains *element theories*. The next level is formed of *theory nets*, while the top level comprises *theory holons* (Balzer, *et al*, 1987). Later it was further developed and applied to different problems in philosophy and history of science by Balzer, Moulines (Balzer, *et al*, 1987), Stegmüller (1976; 1979) and many other researchers.

The most advanced structural representation of scientific knowledge in general and scientific theories in particular is developed in the *structure-nominative direction* in methodology of science which comprises all main directions in methodology of science (Burgin and Kuznetsov, 1991; 1992; 1992a; 1993; 1994; Balzer, *et al*, 1991). According to the structure-nominative model, the structure of a developed scientific or mathematical theory, as well as of the whole system of knowledge, is comprised of five *basic subsystems*: the *logic-linguistic subsystem*, *model-representation subsystem*, *pragmatic-procedural subsystem*, *problem-heuristic subsystem*, and *subsystem of ties* (see Figure 3.13). All these systems are separated and studied based on functions and functioning of real scientific and mathematical theories.

It is natural to call system of knowledge with all five basic subsystems complete knowledge systems and treat minimal knowledge systems as *element knowledge systems*. Note that not only theoretical knowledge has complete knowledge systems and element knowledge systems but also many other areas, e.g., medicine or engineering. Element knowledge systems are elements of knowledge at the megascale, which is the upper scale in the world of knowledge. *Element theories* from the structuralist model (Balzer, *et al*, 1987) and *theory-units* defined in (Burgin, 2011) are examples of element knowledge systems.

Element knowledge systems form larger knowledge systems called global knowledge nets. Namely, when some element knowledge systems are related to, e.g., represent, some domain, then they are connected in a network called a *global knowledge net*. *Theory-nets* from the structuralist model (Balzer, *et al*, 1987) and more general *theory-nets* defined in (Burgin, 2011) are examples of global knowledge nets.

Complexes of global knowledge nets that are tied by synchronic knowledge-related links form knowledge *holons*. Evolution of global knowledge nets through historical time is represented by evolutionary global knowledge systems.

The structural approach from the methodology of science has been also used in several other areas. For instance, Pleiffer and Niehaves (2005) used this approach to evaluate conceptual models in information system engineering.

Chapter 4

STRUCTURALISM AND STRUCTURAL REALISM

In modern science and philosophy, a general approach to cognition based on discovery of basic structures and exploration of their transformations is called *structuralism*. The primary concern of this direction is *structure* rather than essence and nature of studied phenomena, such as, for example, content of texts or their specificities of usage in linguistics. The main assumption of the structural approach is that different things and phenomena have similar essence and many common properties if their structures are identical. In this context, to investigate a phenomenon means to find and study its structure.

As a scientific methodology, *structuralism* is based on the structural vision of reality and on mathematics as the science of structures. It explores the relationships between fundamental principal elements in language, literature, and other fields upon which some higher mental, linguistic, social, or cultural structures and structural networks are built. Through these networks meaning is produced within a particular person, organization, social system, community or culture. This meaning then frames and motivates the actions of individuals and groups.

Structuralism appeared in academia for the first time in the 19th century. Namely, the founding father of experimental psychology Wilhelm Maximilian Wundt (1832-1920) conjectured and tried to confirm experimentally that consciousness can be broken down into fundamental elements, relations between which give birth to complex mental structures. His approach was further developed by his student Edward Bradford Titchener (1867-1927) who worked at Cornell University and used the word *structuralism* to describe this direction in psychology. In spite of this, some think that structuralism was invented in the 20th century when Ferdinand de Saussure applied structural methods to natural languages.

The main idea of Titchener and his followers was that it is possible to understand and analyze perception in terms of elementary *conscious experiences*, such as sweet, sour, cold, warm, bitter, and whatever else could be identified by trained observation (introspection). To form perceptual processes, these elementary conscious experiences were combined in various structures. However, Titchener's structuralism was quickly abandoned because its objects, conscious experiences, were not easily subjected to controlled experimentation in the same way as, for example, behavior is. As a negating reaction to structuralism of Titchener, such psychological directions as *functional psychology* and *behaviorism* appeared. Nevertheless, it is possible to consider ideas of Wundt and Titchener about the role of structures as a root of modern structuralism.

Another root of modern structuralism was *Russian Literary Formalism* (Erlich, 1981). This direction was formed by a group of Russian writers, critics and theorists in the early decades of the 20th century. They assumed and proclaimed that in the art, content simply does not matter, while everything is defined by form being the singular determinant of its status as art. One of the founders of Russian Literary Formalism Viktor Shklovsky (1893-1984) taught his readers principles of Literary Formalism in a series of landmark papers on theories of poetry and the novel.

In a similar way, the well-known French philosopher Jacques Derrida (1930-2004) proclaimed that (in structuralism) the individuality of the text disappears in favor of looking at patterns, systems, and structures. This approach was inherited from the Russian Formalists who proposed that all narratives could be charted as variations on certain basic universal narrative patterns. Inspired by ideas of formalism, Vladimir Propp (1897-1970), in his book *Morphology of the Folktale* (1928), set out to show for a group of 150 Russian folk tales that all of them could be generically classified in terms of thirty one narrative functions distributed among seven dramatis personae. Each story is comprised of narrative functions occurring in an invariant order. This interesting classificatory achievement becomes even more interesting when we understand that Propp's empirically derived classification appears to fit an indefinite number of stories - oral, written, enacted or filmed - and from diverse cultures.

Literary Formalism is also closely linked to New Criticism, the British and American critical movement. For instance, Cleanth Brooks outlines a virtually identical set of conclusions about literary structures and their universality (Brooks, 1947; 1971). He argues that irreducible contradiction (or paradox) is the structural principle that organizes all great works of art. This is one of the basic principles of literary structuralism.

Reappearing in the 20th century, structuralism was associated with analysis of the formal properties of language. It had the form of structural linguistics, which produced a major impact on a number of disciplines. The prominent Swiss linguist Ferdinand de Saussure (1857–1913) was the originator of this reappearance because in his studies, he focused not on the use of language in communication, i.e., on the speech (*parole* in French), but rather on the underlying structure of the language (*langue* in French). He was interested in the infrastructure of language that is common to all speakers and that functions on the unconscious level. Although Saussure did not use the term *structure* preferring the term *form*, his understanding in many respects coincides the principles of structuralism (de Saussure, 1916; 1916a). Saussure called his theory *semiology*.

In the 19th century, the study of languages had been a historical subject, exploring such problems as how languages had evolved over time or how European languages had developed from a common Indo-European matrix. It was the *diachronic approach* to linguistics. In contrast to this, Saussure dropped the historical dimension (a move that would become common in one discipline after another in the 20th century). He examined how the elements of language are related to each other in the present, revealing *synchronic structures* rather than dynamic and changing over time *diachronic structures*.

Saussure's linguistic inquiry was centered on the underlying rules and conventions enabling language to operate. He pioneered and promoted study of grammar rather than usage, rules rather than expressions, how language works rather than what people speak about, models rather than data, language rather than speech. However, his study was done via examination of the speech, expressions and usage.

Saussure's approach was based on three main principles:

1) Language is a structured system with the priority of the whole over the parts.
2) The elements of language function in process of combination and contrast with each other;
3) Linguistic signs have no inherent meaning, but only acquire meaning by virtue of their relations with one another.

Works of de Saussure had a great influence on other linguists, such as Leonard Bloomfield (1887-1949) in America, Antoine Meillet (1866-1936), André Martinet (1908-1999) and Émile Benveniste (1902-1976) in France, Louis Hjelmslev (1899–1965) in Denmark and Alf Sommerfelt (1892-1965) in Norway, who further developed his approach.

The terms *linguistic structure* and the *structural linguistics* were introduced by Rasmus Viggo Brendal (1887-1942) in (Brendal, 1939). Hjelmslev considered structure as an "independent essence with inner dependencies". He emphasized that the properties of each element of linguistic objects depend on their structures (Hjelmslev, 1958).

Another stream of structuralism came from the *Prague School* of linguistics, also known as the Prague Linguistic Circle. That is why the birth of European structuralism is often related to the conference of the Prague School of linguistics in The Hague in 1928. Members of the Prague Linguistic Circle, such as Roman Jakobson (1896-1982) and Nikolai Trubetzkoy (1890-1938), had conducted research that later became very influential.

The main contribution of the Prague School was to phonology. Rather than simply compile and describe sounds occuring in natural languages, linguistics from the Prague School studied how they were related analyzing sounds in terms of a system of phonemic oppositions, such as the sounds /p/ and /b/ or /r/ and /l/ in English. Later Jakobson applied the structural techniques of analysis to syntax, morphology, and semantics of languages. This approach has become the paradigmatic basis for structuralism in a number of different forms.

Now linguists study languages and communication by means of a variety of structures because any developed natural language, such as English or Spanish, comprises a diversity of structures allowing this language to perform its functions. For instance, information structure of sentences is widely used in linguistics (Halliday, 1967; Roberts, 1996; Steedman, 2000). This structure reflects the partitioning of sentences into categories, such as the background, focus, topic, comment, etc.

Structuralism in linguistics and literary studies is appealing to researchers because it adds certain objectivity to the realm of research. This scientific objectivity is achieved by subordinating speech to language and abandoning actual usage in favor of studying the structure of a system in the abstract form. Structuralist readings ignore the specificity of actual texts and treat them as if they are like the patterns produced by some impersonal force or power and not the result of human effort.

Reappearing in the second half of the 20th century, structuralism was soon extended to various other fields, becoming a major post-war intellectual movement in Europe and the United States, especially, in academic fields concerned with the analysis of language, culture, and society. Academic interest began around 1958 and peaked in the late 1960s and early 1970s. Structuralism as a scientific methodology brought the notion of *structure* to a diversity of areas, including sociology, anthropology, esthetics, linguistics, economics, and psychology (Piaget 1971; Gryakalov, 1989).

The first prominent figure in this area was Claude Levi-Strauss (1908-2009) who is often considered to be the first to use the term *structuralism* in the 20th century. He developed and successfully applied the structuralistic methods and structural analysis in anthropology (Lévi-Strauss, 1963; 1987). It is interesting to know that Levi-Strauss spent the war years in exile in New York. There he met the member of the Prague School Roman Jakobson, who transmitted the legacy of structuralist linguistics to Levi-Strauss. Through this decisive influence on Claude Lévi-Strauss and Roland Barthes (1915–1980), among others, Jakobson became a pivotal figure in the adaptation of structural approach to many disciplines beyond linguistics.

Levi-Strauss, following in the footsteps of Jakobson, assumed that the human mind operated in terms of such elementary structures as binary oppositions, which structured all the phenomena of human culture. As a result, he and other structuralists described different conceptual systems, such as the individual or social mental world, as structures built of binary oppositions, such as light/darkness, hot/cold, culture/nature, being/nothingness, true/false and white/black. These oppositions structure meaning and mentality, making possible, for example, to represent and study various fields of cultural thought in terms of these binary oppositions.

We remind that any opposition is a fundamental triad described in Chapter 1 because two concepts form an opposition only there is a specific connection, namely "being opposite," between them. For instance, the following oppositions in the form of a fundamental triad are often considered:

$$\text{light} \xrightarrow{\textit{is opposite to}} \text{darkness}$$

$$\text{good} \xrightarrow{\textit{is opposite to}} \text{bad (evil)}$$

$$\text{hot} \xrightarrow{\textit{is opposite to}} \text{cold}$$

$$\text{culture} \xrightarrow{\textit{is opposite to}} \text{nature}$$

$$\text{being} \xrightarrow{\textit{is opposite to}} \text{nothingness}$$

$$\text{true} \xrightarrow{\textit{is opposite to}} \text{false}$$

$$\text{white} \xrightarrow{\textit{is opposite to}} \text{black}$$

On this basis, Levi-Strauss had restored the common (for all cultures) function that intermediates the fundamental contradictions of the human existence. In his opinion, this function has been lost by the modern European civilization.

Extending the approach of Jakobson and Levi-Strauss, Derrida approaches all texts as constructed around elementary oppositions and all speech has to articulate them if it intends to make any sense whatsoever. He explained that textual constructs only produce meaning through the interplay of differences inside a "system of distinct signs".

Following the tradition of de Saussure, Trubetzkoy and Jakobson, Levi-Strauss specified properties of procedures basic to structuralism:

(1) Structural analysis examines structural aspects of the studied phenomena.
(2) The emphasis is on relations and not on separate elements.
(3) Attention is primarily on the system and not on independent entities.
(4) The goal is to find general laws accounting for the underlying organizing patterns of the studied phenomena.

Levi-Strauss gave rise in France, to the "structuralist movement," the most well-known representatives of which were such thinkers as Jacques Marie Émile Lacan (1901-1981), Michel Foucault (1926–1984), and Louis Althusser (1918–1990). For instance, Foucault (1966/1973) examined the history of science to find how structures of epistemology, that is, theory of scientific cognition, shaped the way in which people comprehended knowledge and knowing.

Structuralism provides means for both the understanding of texts and the understanding of cultures. Through semiotics, the science of signs, structuralism leads researchers to study everything as texts and by means of texts, which are composed of signs, governed by conventions of meaning, and ordered according to patterns of relationships. Treating some object (system) as a text projects a textual structure onto this object (system). The cultural construction is identified in terms of the relations of signs that constitute the meaning and substance of culture. For instance, big amounts of money or vast property are signs (symbols) of success in contemporary society. Some signs carry with them several levels of cultural meaning. Such second-order signifiers (signs) are called, by Barthes, myths. Anything can be a *myth* in this sense. For instance, very expensive watches are a mythic signifier of success.

However, structural anthropology, as well as other directions of structuralism in the humanities, fell out of favor in 1980s for a number of reasons. One of them was unverifiable assumptions made about the universal structures of the human mind. Researchers also argued that political economy should be more at the forefront of anthropology. In addition, anthropologists became more concerned with how cultural and social structures were changed by human agency and practice. Besides, structuralism rejected the concept of human freedom and choice, focusing instead on the way that human behavior is determined by various structures.

In spite of finding considerable fault with Lévi-Strauss's version of structuralism, there are anthropologists who did not turn away from a fundamental structural basis for human culture. For instance, the Biogenetic Structuralism group argued that some kind of structural foundation for culture must exist because all humans inherit the same system of brain structures. They proposed a kind of neuroanthropology which would lay the foundations for a more complete scientific account of cultural similarity and variation by requiring an integration of cultural anthropology and neuroscience.

In contrast to the decline of structuralism in the humanities, structuralism in science, philosophy of science, mathematics and philosophy of mathematics has become a flourishing area attracting more and more adherents. In all these disciplines, structures have been extensively studied and utilized as one of the basic conceptions. For instance, structuralism applied to physics explains that physics provides us with knowledge of the physical structures that exist in the world.

It is possible to find elements of structuralism in the works of such notable philosophers as Ernst Cassirer (1874-1945), Moritz Schlick (1882-1936) who was the originator of logical positivism, Rudolf Carnap (1891-1970), Bertrand Russell (1872-1970) and Grover Edward Maxwell (1918-1981), as well as in ideas of such prominent physicists as Arthur Stanley Eddington (1882-1944) and Hermann Weyl (1885-1955).

The transition to the study of a structural reality also reflects a new stage of the structuralist doctrine. In particular, investigation of the structural level of the world and scientific research of structures as real phenomena alleviated finding the most basic structure called *fundamental triad* or *named set*.

Structuralism in mathematics, or *mathematical structuralism*, is the study of what structures (mathematical objects) are and how the ontology of these structures should be understood. It is rooted in Richard Dedekind's approach to natural numbers (Dedekind, 1887), was clearly formulated by Bertrand Russell (Russell, 1919; 1927) and by Albert Lautman and Nicolas Bourbaki in a more formalized representation (Lautman, 1938; Bourbaki, 1957; 1960). It is a growing philosophy within mathematics that is not without its share of critics.

In addition, philosophical problems of mathematical structuralism have been studied by many philosophers who developed a distinct direction in the philosophy of mathematics (see, for example, (Resnick, 1997; Shapiro, 1997; Hellman, 2001; 2005)).

There are different forms of mathematical structuralism.

Non-eliminative structuralism (NES) assumes that pure mathematics is the investigation of *abstract* mathematical structures, which comprise all mathematical objects.

Eliminative structuralism (ES) takes mathematics to be the investigation of *concrete* (or *embedded*) structures, which are structures of mathematical systems (objects).

Similarly to philosophical views on universals, directions in mathematical structuralism are also classified with respect to the assumption about reality of structures.

Platonic *ante rem structuralism* holds that mathematical structures exist objectively, independently of the mathematicians and are also independent of whether the structures are exemplified in the non-mathematical realm. On this view, structures are ontologically prior to any systems that have them.

Aristotelian *in re structuralism* regards that structures exist, but only in the systems that instantiate them. Destroy all the natural number systems and, alas, you have destroyed the natural number structure itself.

Ockham's *conceptual structuralism* presumes that structures do not exist as objects but only as conceptualized frameworks situated within the thinking mind providing convenient shorthand for talk of systems that have a certain similarity.

There is a general opinion that structuralism re-appeared quite recently in the context of science in general and physics in particular, where it is called structural realism.

There are two kinds of *structural realism*: structural realism in politics and political sciences and structural realism in philosophy.

Structural realism in politics and political sciences is based on the following assumptions:

- The international system is anarchic, that is, there is no credible power above the states controlling the whole system.
- States cannot be certain of the intentions of other states.
- Some states have offensive capabilities.
- States have preferences and survival is a prerequisite for realizing such a preference.

From these premises, *structural realism in politics* comes to the following conclusions:

- As states require survival in order to realize their preferences, they seek to survive.
- As states cannot be certain of the intentions of other states, which may have offensive capabilities, and because there is no higher authority which can protect them from those other states, it is rational for states to seek some optimal level of power relative to all other states in the system. Relative power can then be used as a means of survival, and therefore a means of achieving the state's preferences.
- The result is an international system in which the majority of states compete with one another for relative power.

Structural realism in philosophy is the stage of the development of scientific realism. While scientific realism assumes that all sensible objects, as well as unobservable entities posited by contemporary science, exist in reality, structural realism in philosophy speaks only about structures of objects but not about objects themselves.

It is necessary to remark that some objects considered in science as real were later rejected by subsequent scientific theories. Examples of such objects are flogiston and ether. These situations demonstrated shortcomings of scientific realism and brought forth structural realism.

Structural realism (SR) is the view or a philosophical position in its developed form that people's knowledge about the world is knowledge about structures without telling us about the natures of things underlying the phenomena. In particular, offering true descriptions of reality, scientific theories tell us only about structure of things they study. As the idea of a structure is generally understood in terms of relations between the elements of a system, structuralism and structural realism focus on the relations themselves, rather than on essence and nature of studied phenomena.

Some philosophers write that structural realism was introduced into contemporary philosophy of science by John Worrall in 1989 as a way to eliminate shortcomings of the standard scientific realism. However, according to Frigg and Votsis (2011), the term structural realism was coined much earlier by Grover Maxwell (1968). Adherents of structural realism insist that it would be reasonable not to accept standard scientific realism, which asserts that the best scientific theories correctly describe the nature of the unobservable objects that cause the observe phenomena. Antirealism about science would also be counterproductive. So, the most reasonable is to adopt structural realism, epistemically committing only to the mathematical or structural content of scientific theories. As a result, structural realism has become a very hot topic in the philosophy of science, and many philosophers now view it as the most defensible form of scientific realism.

It is useful to know that several scientists and mathematicians had expressed views that became the essence of structural realism long ago. It is possible to find elements of structural realism in philosophical writings of Poincaré (1902; 1905; 1908) and Duhem (1906). For instance, explaining his views on scientific cognition, Poincaré (1902) wrote:

"The aim of science is not things in themselves, as the dogmatists in their simplicity imagine, but the relations between things; outside those relations there is no reality knowable."

In addition, Poincaré describes the role of science to be a structural one, writing (Poincaré, 1902):

"Science is built up of facts, as a house is built of stones; but an accumulation of facts is no more a science than a heap of stones is a house."

McMullin (1990) argues that Pierre Duhem was a realist about the relations found in laws but not about explanations in terms of an ontology. According to Worrall (1989), Gower (2000) and Zahar (2001), Duhem too was a kind of structural realist, though there are passages in Duhem that more readily lend themselves to an instrumentalist interpretation.

However, in a fully developed form, structural realism first appeared in (Russell, 1912) where it was written:

"... although the relations of physical objects have all sorts of knowable properties, derived from their correspondence with the relations of sense-data, the physical objects themselves remain unknown in their intrinsic nature."

In other words, Russell advocated the view that although there are good reasons to believe that the causes of sensations are physical objects, all what is possible to know about these objects is their structure.

Later he stated (Russell, 1927) that "the only legitimate attitude about the physical world seems to be one of complete agnosticism as regards all but its mathematical properties." This statement expresses the approach of structural realism because mathematical properties represent relations without any other knowledge about the external world.

It is interesting that the noted logician Paul Isaac Bernays (1888-1977) had a similar intuition with respect to axioms regarding them not as a system of statements about a subject matter but as a system of conditions for a relational structure. He wrote (Bernays, 1967):

"A main feature of Hilbert's axiomatization of geometry is that the axiomatic method is presented and practiced in the spirit of the abstract conception of mathematics that arose at the end of the nineteenth century and which has generally been adopted in modern mathematics. It consists in abstracting from the intuitive meaning of the terms ... and in understanding the assertions (theorems) of the axiomatized theory in a hypothetical sense, that is, as holding true for any interpretation ... for which the axioms are satisfied. Thus, an axiom system is regarded not as a system of statements about a subject matter but as a system of conditions for what might be called a relational structure ... [On] this conception of axiomatics, ... logical reasoning on the basis of the axioms is used not merely as a means of assisting intuition in the study of spatial figures; rather, logical dependencies are considered for their own sake, and it is insisted that in reasoning we should rely only on those properties of a figure that either are explicitly assumed or follow logically from the assumptions and axioms."

Some philosophers had also expressed views that became the essence of structural realism long before the term appeared. For instance, Rudolf Carnap (1928) stated that all scientific statements may be reducible to structural statements about basic elements of experience. Structural statements were logical propositions about the formal properties of objects and relations. Carnap (1928) also explained that the formal properties of objects and relations between them, and not their material properties, determined their applicability to a logical construction of the world.

Contemporary structural realism comes in two basic forms: epistemic structural realism and ontic structural realism.

Ontic, also called *eliminative* or *ontological, structural realism* (OSR) holds that we can know structural aspects of reality, and that there is nothing else to know, meaning that objects are useful heuristic constructs, but are ultimately ontologically dispensable. In other words, some philosophers postulate an ontology of structures. This form of structural realism implies that not only subatomic particles or gravitational field but also what we see in our mundane life, for example, trees, tables, chairs and cars, do not exist as things but persist only as (conventional) structures.

Epistemic, also called *restrictive* or *epistmological*, structural realism (ESR) holds that we can know structural aspects of reality, but nothing about the natures of studied things relations of which define structures (Worrall, 1989; Zahar, 1996). In other words, the 'nature' of the underlying objects remains hidden.

Epistemic structural realism has two forms: direct epistemic structural realism and indirect epistemic structural realism.

Direct epistemic structural realism (DESR) holds that we can have fully-fledged knowledge (i.e. structural and non-structural knowledge) about the observable parts and aspects of the world (thus understood), but that our knowledge of the unobservable is only structural.

Indirect epistemic structural realism (IESR) is a descendant of indirect realism, the position that we only have direct epistemic access to our sense data, our perceptions or at any rate something sensory, and only through them do we have (indirect) access to the world itself. For this reason not only neutrinos and the like, but also chairs, tables and rabbits count as unobservable, and claims about the external world tout court can be only structural.

To conclude this chapter, it is necessary to mention *semirealism* of Anjan Chakravartty, who argues that scientific realism as the knowledge about objects implies structural realism as the knowledge about structures and vice versa. According to him, structuralism is actually a form of scientific realism and he calls this form by the name *semirealism* (Chakravartty, 1998; 2003; 2004; 2007).

Chapter 5

WHAT IS REAL: DIMENSIONS OF REALITY AND LAYERS OF EXISTENCE

The problem what really exists or what reality is has been in the center of philosophy from the very beginning of its existence. Philosophers suggested different approaches to this problem and created theoretical systems and philosophical directions based on these approaches. In Western philosophy, the main of these directions are: *idealism*, *materialism*, *realism*, *pragmatism* and *existentialism*.

Idealism is a philosophical approach according to which ideas or forms are the only true reality, while the world of sensible things is an imperfect reflection of this world of ideas/forms. This world is comprehensible only by conscious reasoning in the mind. This approach often called *Platonism* was originally elaborated by Plato (429-347 BCE). Extreme forms of idealism reject existence of anything beyond the world of ideas or forms. For instance, some philosophers, e.g., such prominent philosophers as George Berkeley (1685 – 1753) and David Hume (1711-1776), as well as adherents of some religious systems, e.g., Buddhists, argue that everything that people perceive by their senses is an illusion. However, there are other approaches and in general philosophers discern several kinds of idealism.

Objective idealism interprets the spiritual, mental or structural as a reality existing outside and independent of human consciousness. Objective idealism asserts that the reality combines and transcends the realities of the object experienced and of the mind of the observer. Platonism is the most explicit form of objective idealism. Such prominent philosophers as Friedrich Wilhelm Joseph Schelling (1775–1854) and Georg Wilhelm Friedrich Hegel (1770–1831) had some forms of objective idealism.

Many mathematicians and scientists were adherents of objective idealism. For instance, Poincaré (1905) wrote:

"Does the harmony the human intelligence thinks it discovers in nature exist outside of this intelligence? No, beyond all doubt, a reality completely independent of the mind which conceives it, sees or feels it, is an impossibility. A world as exterior as that, even if it existed, would for us be forever inaccessible. But what we call objective reality is, in the last analysis, what is common to many thinking beings, and could be common to all; this common part, we shall see, can only be the harmony expressed by mathematical laws. It is this harmony then which is the sole objective reality, the only truth we can attain ..."

This is a clear explanation that knowledge of a mind-independent reality is, for Poincaré, impossible, and a demonstration of his epistemic idealism. It is objective as it demands inter-subjective agreement between human beings, or beings with "sufficiently similar" faculties.

Subjective idealism claims that the reality is formed by the mind of an individual. For instance, Berkeley affirmed that the objects of perception are actually ideas in the mind and reality is a mental construct.

Pluralistic idealism takes the view that there are many individual minds that together underlie the existence of the observed world and make possible the existence of the physical universe. It is a version of socially created reality. The great mathematician and philosopher Gottfried Wilhelm Leibniz (1646-1716) had some form of pluralistic idealism.

The great philosopher Immanuel Kant (1724-1804) introduced *transcendental idealism*, which maintains that the mind shapes the world people perceive into the form of space-and-time.

Realism, which can be considered as a kind of idealism, asserts that there is a reality independent of any beliefs, perceptions and/or arguments. Generally, when it is assumed that it is possible to identify a class of objects, the existence or essential characteristics of which do not depend on perceptions, beliefs, language, or any other human artifacts, philosophers speak of "realism about" the objects from that class.

Materialism holds that physical matter is the only or fundamental reality, which exists independent of the human mind, while all being, processes and phenomena (including consciousness) can be explained as manifestations or results of material interactions. Aristotle (384-322 BCE), a student of Plato who broke with his mentor's idealist philosophy, is often called the father of materialism.

The central tenet of *Pragmatism* is that only those things that are experienced or observed are real. Pragmatists believe that reality is constantly changing and that we learn best through applying our experiences and thoughts to problems, as they arise. Pragmatism is rooted in the teaching of the prominent philosopher and semiotician Charles Sanders Peirce (1839-1914), who believed that thought must produce action, rather than linger in the mind and lead to indecisiveness. It was mostly developed by the noted philosopher John Dewey (1859-1952).

According to *Existentialism* the nature of reality is subjective, and lies within the individual. The material world has no inherent meaning outside of human existence. Individual choice and individual standards rather than external standards are central. People define ourselves in relationship to that existence by the choices we make. The early 19th century philosopher Søren Kierkegaard (1813 –1855) is widely regarded as the father of existentialism.

The view that reality is merely a social, or cultural, artifact, is called *social constructionism*. *Cultural relativism* is the view that social issues such as morality are not absolute, but at least partially cultural artifact.

In contrast to the classical schema, in which everything is evaluated in the binary scale with only two possible values – "exists" ("is real") and "does not exist" ("is not real"), our approach suggests more gradations of reality and dimensions of existence.

At first, we consider three pure forms of reality:

1. *Actual reality* consists of natural objects and processes directly or indirectly perceived by senses and reflected in the central nervous system (CNS).

2. *Virtual reality*, often called *virtuality*, is created, simulated or reflected by some technological system (device or machine), e.g., computer games, movies and videos.
3. *Imaginary reality* is created by mentality, e.g., heroes in prose and poems or characters in movies and plays.

Now *virtual reality* (VR) is a very popular term because virtual reality technology is used in a variety of fields from industrial to military to medicine to entertainment. On the Internet, the abbreviation RL stands for "real life" and IRL means "in real life". These terms are used to discern actual reality from virtuality. In spite of the VR ubiquity, different authors suggest and use different definitions of virtual reality. As a result, there is no definition accepted by all researchers and practitioners. Comparing all existing definitions, we see that the above definition is the most general. As an example of a different definition, we can take the following one:

Virtual reality (VR) often denotes computer-simulated environments that can simulate physical presence in places in the real world, as well as in imaginary worlds.

The *cyberspace*, the world that exists in computer systems considered as an interconnected whole, is thought of as a virtual reality, which consists of a diversity of virtual worlds.

Many trace virtual reality to the 1860s, when 360-degree art through panoramic murals began to appear. However, according to our definition, virtual reality exists since that time when people started to use devices that created images perceived by senses. The origin of the term virtual reality is usually attributed to the French playwright, poet, actor, and director Antonin Artaud (1938).

It is possible to consider three cases of virtual reality: *substitutional* or immersive virtual *reality*, *interactional virtual reality* and *observational virtual reality*.

Substitutional virtual reality substitutes actual reality for a person who is immersed in this virtual world. For instance, in brain-computer interface simulations, participant enter from outside by directly connecting their brains to the simulation computer. The computer transmits sensory data to the participants and back, reading and responding to their desires and actions. In this manner, the participants interact with the simulated world and receive feedback from it. The participants may be induced by any number of possible means to forget, temporarily or otherwise, that they are inside a virtual realm. While inside the simulation, the participant's consciousness is represented by an avatar, which can look very different from the participant's actual appearance.

This *virtual* or *simulated reality* is the result of computer simulation of real processes to a degree indistinguishable from actual reality. It could contain conscious minds, which may or may not be fully aware that they are living inside a simulation. However, this level of virtuality (simulation) is quite different from the current technological level where virtual reality is easily distinguishable from actual reality. The participants are never in doubt about the nature of what they experience. In future, with the development of technology, it will be hard or impossible to separate simulated reality from actual reality.

Such a prognosis inspired several science fiction writers to suggest that people and their world do not exist as physical systems but are created and simulated by some supercomputer existing in the world developed and sustained as a virtual reality by the same supercomputer. There has been much debate over this topic, ranging from philosophical discourse to practical applications in computing.

Interactional virtual reality allows interaction of people and physical systems with objects and processes created and supported by technical devices. The most popular examples are computer games, training simulators and educational computer programs. However, even display of any software system such as a word processor or operating system also is an interactional virtual reality.

Observational virtual reality allows only observation (watching) of objects and processes created and supported by special devices. Movies and videos on the TV screen are examples of observational virtual reality.

Examples of *imaginary reality* are *symbolic universes* introduced and studied by Berger and Luckmann (1966). *Symbolic universes* are sets of beliefs on the level "everybody knows." They aim at making the institutionalized structure plausible for the individual (who might not understand/agree with the underlying logic of the institution). These universes are created to provide legitimation to the created institutional structure. A symbolic universe "puts everything in its right place" in the social life. It provides explanations for why people do things the way they do. Proverbs, moral maxims, wise sayings, mythology, religions and other theological thought, metaphysical traditions and other value systems are parts of symbolic universes. All symbolic universes form the *symbolosphere* (Burgin and Schumann, 2006).

Symbolic universes exist on different levels – from society as a whole through social institutions, such as religion, state, education or science, through organizations, such as universities, companies or clubs.

There are also four combined forms of reality:

- *Mixed reality* is a combination of actual and virtual reality, i.e., it is situated between actual reality and virtual reality forming the *Virtuality Continuum* and including *augmented reality* and *augmented virtuality.*
- *Materialized reality* is a combination of imaginary and virtual reality.
- *Actualized reality* is a combination of imaginary and actual reality.
- *Enhanced reality* is a combination of actual, imaginary and virtual reality.

Object that are perceived by means of some technical devices, such as telescopes, microscopes, TV transmitters and receivers, belong to the mixed (augmented) reality. The more sophisticated devices are for perception, the more augmented reality is. All this shows that there are grades and extents of virtuality added to the actual reality. These grades and extents represent the *Virtuality Continuum*, the term used to describe the concept that there is a continuous scale, which ranges between the completely virtual reality and actual reality. The reality-virtuality continuum therefore encompasses all possible variations and compositions of real and virtual objects. It has been described as a concept in new media and computer science, but in fact it could be considered a matter of anthropology. The concept was first introduced in (Milgram and Kishino, 1994).

In *mixed reality*, people and other physical systems coexist and interact in real time with digital objects. As an intermediate case in the *Virtuality Continuum*, it refers to predominantly virtual spaces, where physical elements, e.g. physical objects or people, are dynamically integrated into, and can interact with the virtual world in real-time. This integration is achieved with the use of various techniques. Often streaming video from physical spaces, e.g. via webcam, or using 3-dimensional digitalisation of physical objects.

In *augmented reality* (AR), the virtual augments the real, while in *augmented virtuality* (AV), the real augments the virtual. Now when people constantly use cell phones, computers, television, and many other technical devices that create virtual reality, people live in augmented reality. Our world has become such an augmented reality.

As an example of *augmented virtuality* (AV), let us consider image-guided neurosurgery described in (Paul, et al, 2005). According to surgeons, displaying anatomical and physiological information derived from preoperative medical images in the operating room is critical in image-guided neurosurgery. To improve performance during the surgery, a calibrated stereovision system was set up between the surgical microscope and the binocular tubes. Then three-dimensional surface images of the operative field were generated. All these images were combined together and augmented with the presence of the patient in the physical space. This created augmented virtuality, which provided better conditions for the surgery.

One more place in the *Virtuality Continuum* is occupied by *computer-mediated reality*, which means processes where computers add, subtract and transform information, in this and other ways manipulating people's perception, such as visual perception of the environment. Usually wearable computers or advanced handheld devices, such as smartphones, are used for this purpose.

Everything that is presented and goes on in non-documentary movies is *materialized reality*. Avatars in computer games are materialized reality. When a non-documentary movie reflects real events, such as, for example, historic chronicles of Shakespeare, the events, heroes and situations in these movies come from the mental world of the playwright where the real events have been reflected and very often transformed. Thus, such movies present the materialized reality.

Plays that are presented and go on the stage of a theater belong to *actualized reality*. Even if a play reflects real events like historic chronicles, such as were written by Shakespeare, they come from the mental world of the playwright where the real events have been reflected and very often transformed.

A higher level of the actualized reality on the Earth is formed by the *sociosphere* considered as the part of a planet's terrestrial system in which social relations and institutions develop and where social interaction of people goes on. It includes political and economical systems. The sociosphere exists as a synthesis of the *biosphere*, which is the global ecological system integrating all living beings, their relationships, and their environment, and *symbolosphere*, which is a component of our world in which symbols emerge, symbolic systems develop, function, and interact and where symbolic interaction of people goes on (Burgin and Schumann, 2006).

The *actualized social reality* is formed in the process of the social construction of reality (Berger and Luckmann, 1966). For instance, social values are constructed through several concomitant sociological processes, such as externalization, reification, and internalization. These social processes constitute reality-loops and depend on the culture. Thus, what is considered to be real in one culture may not be so in another.

What is socially constructed often is considered to be real. As a result, the past is always undergoing redefinition. In this context, language becomes a medium for reality construction because ideas, thoughts, and feelings cannot be shared with others unless they are first materialized through language (Berger and Luckman, 1996). Only using language people learn how to define the past, understand the present, and dwell on the future. All this constitutes the pragmatics of language considered in Chapter 7.

Such phenomena as the social construction of reality bring forth the *Imagery Continuum* describing the concept that there is a continuous scale, which ranges between the completely imaginary reality and actual reality. For instance, it includes *augmented imagery*, in which people add real events to what they only imagine creating in such a way new reality for themselves.

A combination of the Imagery Continuum and the Virtuality Continuum brings us to the *Enhaced Continuum* where actual, imaginary and virtual are intermingled. In the contemporary society based on advanced technology many people exist in this Continuum.

Having combined forms of reality, we can introduce measures of their belonging to one of three pure cases: a *measure of actuality*, a *measure of virtuality* and a *measure of imagery*. These measures determine the place of reality in the *Imagery Continuum*, *Virtuality Continuum* and *Enhanced Continuum*.

In addition, each of the pure forms of reality is further subdivided into three subforms according to their correspondence features:

1. Actual reality can be: *true* when the reflection is correct, *distorted* when the reflection is partially correct, and *false* when the reflection is incorrect. If a false or distorted reality is comprehended as true, it is called illusion.
2. Virtual reality can be: *possible*, *correlated/similar*, and *impossible*.
3. Imaginary reality can be: *feasible, matching, and unfeasible.*

The whole schema of reality forms and subforms is presented in Figure 5.1.

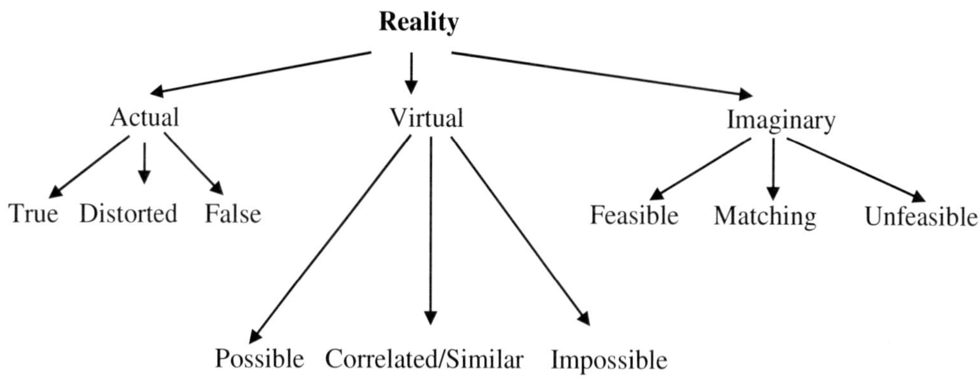

Figure 5.1. Reality structuration.

Forms of existence constitute the *medium plane* of existence with two dimensions – virtual and imaginary. The medium plane has the complementary, *actualization dimension*. All three dimensions are presented in Figure 5.2, forming the medium space.

These dimensions are similar to the dimensions of Hartmann (1940; 1952), who regarded three ontological dimensions: *aspects* or *moments of being*, *spheres of being*, and *modalities of being*. The scale of aspects/moments of being has two elements - *Dasein* and *Sosein*, or *that* and *what*. The scale of spheres of being has four elements/spheres – two primary ones, which are also designated as modes of being (*Seinsweisen*): *real*, which consists of the chain of temporal events, and *ideal*, and two secondary ones, *cognitive* and *logical*. The scale of the modalities of being has three elements/modes - *actuality, possibility*, and *necessity*. There are important relations between the primary and secondary spheres: an intimate cross-relationship between the cognitive and the real modes of being and similar relationship between the logical and the ideal.

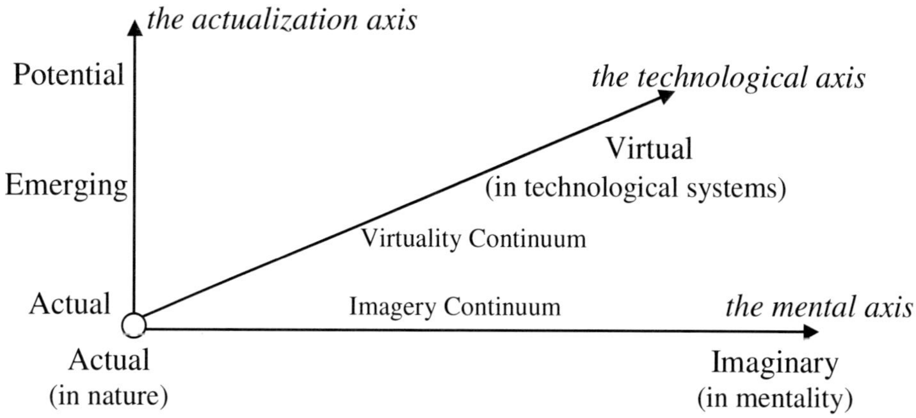

Figure 5.2. Reality dimensions.

Potential aspects of reality were discussed by the great physicist Werner Heisenberg who wrote (see (Herbert, 1885)):

"In the experiments about atomic events we have to do with things and facts, with phenomena that are just as real as any phenomena in daily life. But atoms and the elementary particles themselves are not as real; they form a world of potentialities or possibilities rather than one of things or facts ... The probability wave ... mean tendency for something. It's a quantitative version of the old concept of *potentia* from Aristotle's philosophy. It introduces something standing in the middle between the idea of an event and the actual event, a strange kind of physical reality just in the middle between possibility and reality."

Forms of existence have different scales. Here we consider some of them.

The *temporal scale* is based on time, indicating when something existed, exists or will exist. For instance, what existed 3000 years ago is not as real, i.e., real to the same extent, as what existed 30 years ago and people still live who witnessed it. At the same time, what existed 30 years ago is not as real as what exists now.

Usually things that will only exist in the future have much less degree of existence than things that exist now. However, existence relative to an individual depends on knowledge and positions of future and present things on the temporal scale may be reversed. For instance, if you are sure that something will exist in the future but do not know about something else that exists now, then the first thing (essence) has higher actuality than the second one.

The *conditional scale* is based on the scope of relevant tolerable conditions for existence. For instance, *extinct species* are less real than *critically endangered species*, which, in turn, are less real than *vulnerable species*. According to the *IUCN Red List of Threatened Species*, the extinct species *Mauritius Blue-Pigeon* was first described in 1602, and persisted for more than two centuries thereafter. It was reported in 1832 that birds were "still found towards the centre of the island in the middle of those fine forests which by their remoteness, have escaped the devestating axe". The last specimen was collected in 1826. Now Mauritius Blue-Pigeon is known only from three skins and a number of descriptions and paintings.

At the same time, according to the *IUCN Red List of Threatened Species*, the critically endangered species *Nightingale Reed-Warbler* is historically known from Guam (USA), Saipan, Alamagan, Aguijan and Pagan in the Northern Mariana Islands (USA). It was extirpated from Guam in the late 1960s, and from Pagan before 1981. A tiny population perhaps survives on uninhabited Aguijan (1-6 birds) and on Alamagan (around 346 birds), but the majority of the population occurs on Saipan.

This species qualifies as Critically Endangered because the very rapid rate of decline in its very small global population observed over the past three generations is expected to increase owing to habitat loss and degradation combined with the impact of brown tree snake, which has already led to the reed-warbler's extirpation from Guam.

One more reality scale is the *experiential scale* calibrated by types of individuals or groups whose experience informs us about existence of objects. Indeed, what you saw is more real to you than what your friend saw and told you about. Based on this scale the renowned philosopher Rudolph Carnap (1928) separated two classes of objects, calling objects of a person's own experience by the name *autopsychological* (*eigenpsychisch*) and objects of others' experience by the name *heteropsychological* (*frempsychisch*). Note that these classes are individual and as a rule, different for different people.

Autopsychological and heteropsychological objects may be objects of subjective experience, but when their formal properties are investigated, then they may also become objects of intersubjective experience. For instance, scientific investigation of the formal properties of objects and relations may transform subjective experience into intersubjective experience, and may transform intersubjective experience into objective knowledge. Carnap explains that intersubjective correspondence of autopsychological and heteropsychological objects may occur if the objects of a person's own experience correspond to the objects of others' experience. Thus, the intersubjective object domain is a unified domain of the autopsychological and the heteropsychological. This unity of the object domain is an important aspect of the logical unity of a constructional system of reality.

The *experiential scale* is supplemented by the *reliability scale*, which reflects the reliability of the sources of information about existence. For instance, when a well-known expert in environmental problems informs you about the global warming it is much more reliable than receiving the same information from a truck driver. However, it is necessary to admit that the reliability scale is rather shaky as people are prone to mistakes, as well as to forged or inflated authority.

In addition, it is possible to consider individual reality, that is, reality for a given system, for example, for a woman or a cat. Taking a system R, we know that with respect to R, there are *inner reality* and *outer/external reality*. Inner reality includes some parts of imaginary reality but can also contain other components. For instance, when R is a physical system, then processes that go in this system belong to its inner reality and are not imaginary but actual.

The reader can ask whether information we get from the environment allows us to obtain objective structures of objects, i.e., to explicate structures that exist in reality, or the structure of a system depends only on people or other systems who build the representation. At the first glance, the process of structure explication looks very subjective as it depends on how it is possible to distinguish elements and how ties are defined. However, a similar situation exists with what people see, hear and describe. For instance, in the situation when an individual sees a table, we can ask the question: *Is it an image of a real table, an image on a screen or only an illusion of the brain*?

Idealists argue that all things that people perceive by their senses is an illusion. Now some philosophers, scientists, and science fiction writers have further developed this idea and suggested possible technical mechanisms that could be used for sustaining such an illusory world. For instance, Rucker (1987) suggests that for postmodern people, reality is a pattern of information. Tegmark (2008) suggests that the external physical reality is a mathematical structure. Mahoney (2010) asserts that science is not about nature, but about how we represent nature to ourselves. According to him, we know about nature through the models we build of it, constructing them by abstraction from our experience, manipulating them physically or conceptually, and testing their implications back against nature. The most developed models have the form of mathematical structures. However, any model (even a physical one) includes some structure.

In contrast to this, science developed means to discern objective reality from illusions. These means are based on repetition of observations and experiments. Although these methods are not perfect, they provide for the whole existence of science and development of the human civilization based on science as a source of the technological progress. Application of scientific methods allows scientists to conclude that structures of real systems, which are called *embedded structures* (see Appendix), belong to the actual reality, while abstract structures belong to the virtual or/and imaginary reality. In contrast to this, according to the Platonic approach, there are *ideal structures*, which exist in a specific world inaccessible to senses, while physical structures are imperfect reflections of the ideal structures. There is also a possibility to assume that ideal structures exist in the mental world.

If we say that structures exist only embodied in material things, then we have to admit that material things exist only in a structured form, i.e., matter (physical entities) cannot exist independently of structures. For instance, atomic structure influences how the atoms are bonded together, which in turn helps one to categorize materials as metals, ceramics, and polymers and permits us to draw some general conclusions concerning the mechanical properties and physical behavior of these three classes of materials. Even chaos and vacuum have their structures.

Knowledge about different forms and bearings of reality allow us to suggest a positive solution to the controversy between Plato and Aristotle about existence of *forms* (*eidos*), which can be interpreted as structures according to the contemporary approach.

The basic contradiction between Plato and Aristotle was related to the independent reality of *forms/structures*. Plato postulated their independent existence, while Aristotle asserted that *forms/structures* existed only in material things.

A possible resolution of this contradiction contends that in material things, *forms/structures* exist in actual reality, while *forms/structures* in the sense of Plato exist in potential reality. This understanding well correlates with the Plato's explanation that the "forms which enter into and go out of material (comprehensible by senses) things are the likenesses of real existences modelled after their patterns in a wonderful and inexplicable manner ..." . This potentiality can be comprehended only by intelligence.

Note that when used in the sense Plato, the word *form* (or the word *idea*) is often capitalized, i.e., it is written as *Form* (*Idea*).This brings us to the *World of Forms/Ideas* (see Chapter 7).

In any case, there are many structures that belong to actual reality. Indeed, even if we agree that structures exist only embodied in material things, then we have to admit that material things exist only in a structured form, i.e., matter (physical entities) cannot exist independently of structures. For instance, atomic structure influences how the atoms are bonded together, which in turn helps one to categorize materials as metals, ceramics, and polymers and permits us to draw some general conclusions concerning the mechanical properties and physical behavior of these three classes of materials. Even chaos has its structure and not a single one.

Moreover, Aristotle considered substance as *potentiality* of things, which do not exist without form, and form (structure) as *actuality* of things, defining substance of a thing as "that out of which" a thing is made or as the "constituents" of that thing. As a result, substance in the sense of Aristotle is not necessarily physical. For instance, letters are the substance of words. Aristotle even calls the parts of a geometrical shape, such as segments, rectangles or triangles, by the name *intelligible substance*. Besides, a thing can both *have* substance and also *be* substance. For instance, clay is the substance of bricks, but bricks, in turn, are the substance of a house. So bricks both have substance (clay) and are substance (for a house).

Chapter 6

SIGNS, SYMBOLS AND CONCEPTS

Structures are represented by signs. Thus, to better understand structures, we consider conceptual representations and theoretical models of the concept *sign*.

There are two main interpretations of the word sign: (1) sign as a physical object with some meaning, and (2) sign as a conceptual (theoretical) structure. We will call a sign by the name *material sign* when we have in mind the first interpretation and by the name *conceptual sign* when we bear in mind the second interpretation.

Examples of material signs are digits 0, 1, 2, 3, 4, 5, 6, 7, 8, 9 used in everyday calculations or letters, such as *a*, *b*, and *c*, from alphabets of natural languages. Icons that we can see on the screen of a computer are also material signs. Here we are mostly interested in conceptual signs, treating material signs as names of conceptual signs.

A theoretical explication of the structure of conceptual signs has been developed in *semiotics* as a general theory of signs. Semiotics studies structure of conceptual signs and their communicative function. In what follows, the word *sign* denotes a conceptual sign, while material signs are used as names of conceptual signs.

As signs exist in a huge diversity of situations, the founder of semiotics, Charles Sanders Peirce (1839-1914) and his follower Charles William Morris (1901-1979) defined semiotics very broadly. They predicted that it would influence a variety of disciplines. For instance, Morris wrote (1938):

"The sciences must look to semiotic for the concepts and general principles relevant to their own problems of sign analysis. Semiotic is not merely a science among other sciences but an organon or instrument to all sciences."

Indeed, semiotics has become an important tool in communication research, information theory, linguistics and the fine arts, as well as in psychology, sociology and esthetics. At the same time, we have the following situation. Although many other disciplines recognize the potential importance of semiotic paradigms for their fields, they have not yet found a satisfying way of integrating them into their domain.

It is possible to find the first theoretical approach to the concept *sign* in works of Ferdinand de Saussure, who is sometimes called the father of theoretical linguistics. Saussure studied *linguistic signs* and according to Saussure (1916), the basic property of a sign is that it points to something different than itself, transcendent to it. To represent this property, Saussure introduced a structural model of *sign* in the form of the *dyadic sign triad* (see Figure 6.1).

signification
signifier ⟶ **signified**

Figure 6.1. The dyadic sign triad of Saussure.

As in many other cases, this triad is a kind of a fundamental triad (named set) described in Chapter 1.

In this triad, a *signifier*, also called *signal* or *signifiant*, may be understood as the sound or written pattern of the word or in actual, physical realization as part of a speech act. At the same time, a *signified*, also called *signifie*, may be treated as the conception or meaning of the *signifier*. According to contemporary views, human reality is a construction and the product of signifying activities, which are culturally specific, culturally determined and often unconscious.

According to Saussure, signs can exist only in relation to other signs. In this context, a linguistic sign is not a link between a thing and a word, but between a concept and a sound/written pattern, where the pattern is the hearer/reader's psychological impression given to him by the evidence of her/his senses. This was rather different from previous approaches in linguistics, which were focused on the relationship between words and the things they designate.

A dyadic model of sign was also supported by Eco (1976), who defined a sign as anything which may be interpreted as standing for something else. This understanding is represented by the following diagram, which is a structural model of *sign*:

sign ⟶ **object**
(sign name)

Figure 6.2. The dyadic sign triad of Eco.

However, in a more detailed model adopted from Hjelmslev (1963), Eco explained that any given sign must be an element of an *expression-plane*, and must therefore be conventionally correlated to one or more elements of a *content-plane*.

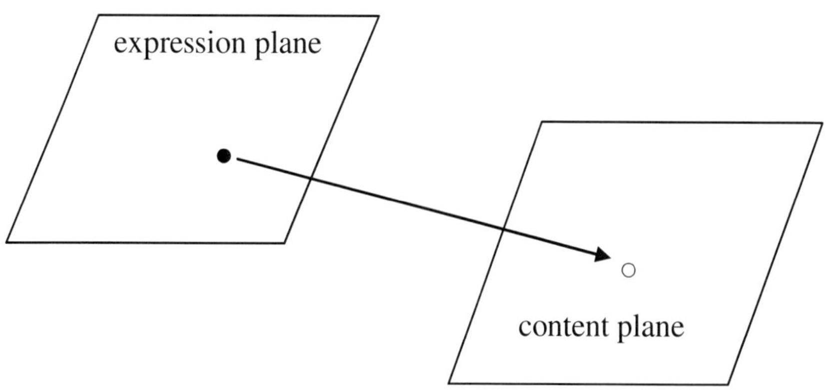

Figure 6.3. The Hjelmslev-Eco model of sign with one expression and one content.

Eco also assumed that it was possible for the expression of a sign to have more than one content (see Figure 6.4).

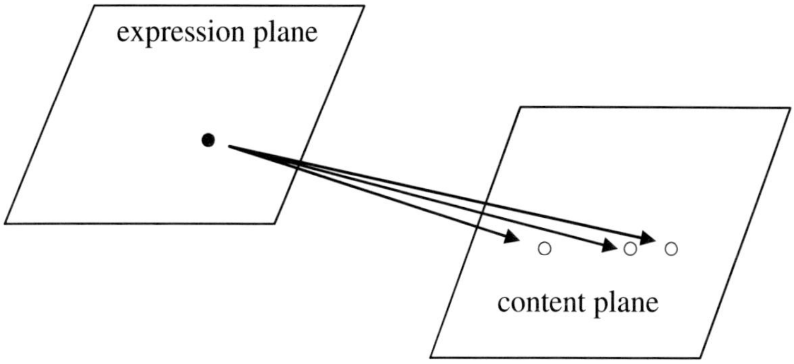

Figure 6.4. The Hjelmslev-Eco model of sign with one expression and several contents.

At the same time, the content of a sign may have more than one expression (see Figure 6.5).

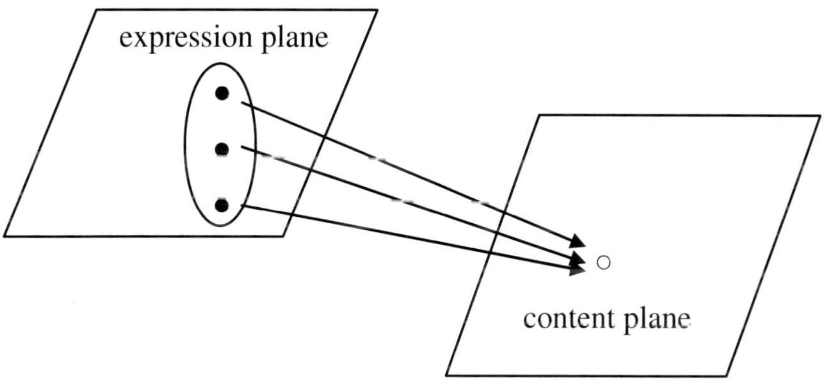

Figure 6.5. The Hjelmslev-Eco model of sign with several expressions and one content.

Besides, in some cases, a sign may have expressions in more than one expression plane, and may have expressions in more than one content plane (see Figure 6.6).

Peirce extended the dyadic Saussure's model by further splitting the signified into essentially different parts: the sign's object and interpretant (meaning of the sign), and thus, coming to the triadic model of a sign, the *balanced sign triad* (see Figure 6.7).

Thus, in the Peirce's model, a sign is understood as a structure that consists of three elements: the Name (Peirce called it Sign Vehicle), Object and Meaning of the sign (Peirce, 1931-1935). Sign Vehicle is conceived as a physical representation of a sign, that is, a material sign, which is called simply *sign* in everyday life. In other words, a sign is often comprehended as some elementary image inscribed on paper, clay tablet, piece of wood or stone, presented on the screen of a computer monitor, and so on.

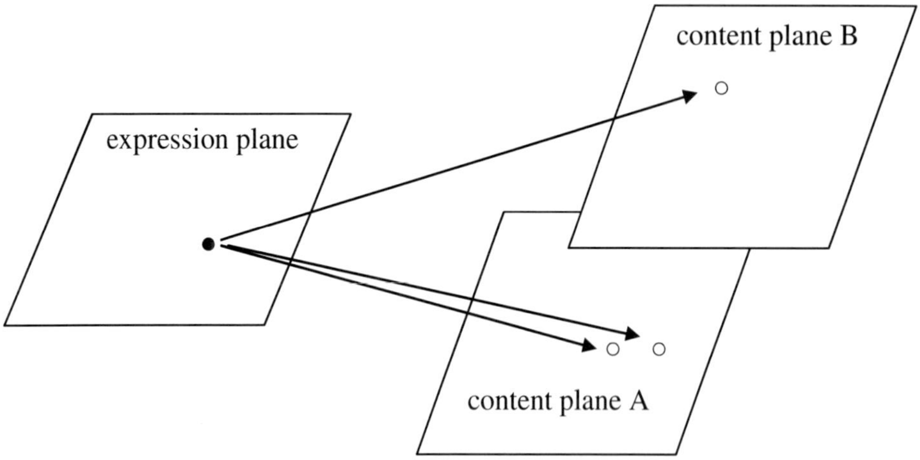

Figure 6.6. The Hjelmslev-Eco model of sign with one expression plane and two content planes.

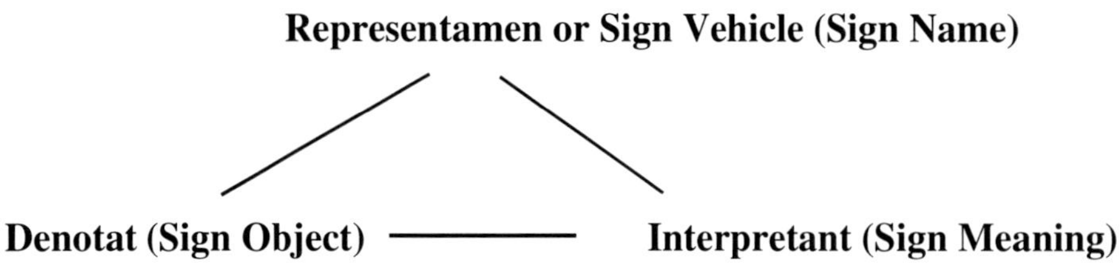

Figure 6.7. The *Balanced Sign Triad* of Peirce.

Peirce implied that signs establish meaning through recursive relationships that arise in sets of three main semiotic elements:

- *Representamen*, also called *Sign* or *Sign Vehicle* by Peirce or *Sign Name* in a general context, is the component of the sign that represents the denoted object or objects and is similar to Saussure's *signifier*. Note that in this context, the sign name is not necessarily a single word. It can be a quite elaborated object.
- *Object*, also called *extent*, is what the sign represents or encodes.
- *Interpretant* is the meaning formed into a further sign by interpreting or decoding a sign.

The object of a sign can be anything thinkable, for example, a law, fact, possibility or idea. Peirce considered two kinds of sign objects:

- The *immediate object* is represented in the sign name (representamen).
- The *dynamic object* is the object as it really is.

In addition, Peirce considered three kinds of the interpretants of a sign:

- The *immediate interpretant* is the meaning that is already in the sign, or more exactly, the meaning that is without delay ascribed to the sign by the interpreter when she/he receives the sign.
- The *dynamic interpretant* is the meaning as formed in a process of sign comprehension by the interpreter.
- The *final interpretant* is the meaning that would be reached if formation process were to be pushed far enough, namely, it is a kind of an ideal meaning, with which an actual, that is, dynamic, interpretant may, at most, coincide.

Note that it also possible to consider the *final* object, which is an ideal collection of the object with all its possible changes.

Besides, as the sign name, representamen or sign vehicle is also an object, it is possible to consider the *immediate sign name*, *dynamic sign name* and *final sign name*.

Peirce explained that signs mediate the relationship between their objects and their interpretants in a triadic mental or mind-like process, which goes through three stages: firstness, secondtness, and thirdness. *Firstness* is a universal category of phenomena and is associated with a vague state of mind in which there is awareness of the environment, a prevailing emotion, and a sense of the possibilities when the mind is in neutral, waiting to formulate thought. *Secondness* is a category associated with moving from possibility to greater certainty expressed by actions, reactions, causality, or reality when the mind identifies what message is to be communicated. *Thirdness* is the category of signs and is associated with generality, representation, continuity, and purpose.

This process is reversed in the receiver. At first, the active mind receives the *sign name* as a material sign. Then the mind acquires from the memory the mental *object* associated with the sign and in such a way produces the *interpretant* of the sign, which is the result of *signification*. Even when a sign represents by a resemblance or factual connection independent of interpretation, the sign is a sign only insofar as it is, at least, potentially interpretable. Peirce also refers to the "ground" of a sign, which is the pure abstraction of the sign quality.

It is interesting that if we analyze what Peirce wrote about his model of sign, we see that he described the structure not of a triangular triad but of a fundamental triad described in Chapter 1. Indeed, Peirce wrote (1903)

"A *Sign*, or *Representamen*, is a First which stands in such a genuine triadic relation to a Second, called its *Object*, as to be capable of determining a Third, called its *Interpretant*, ..."

This gives us the following diagram, which is a fundamental triad because the Object connects the Representamen with the Interpretant.

Representamen ⟶ Object ⟶ Interpretant

Being a thorough taxonomist, Peirce offered a taxonomic scheme of signs demonstrating there could be 59,049 types of signs. However, on the top level of this classification, according to Peirce, there were three basic types of signs: *symbol*, *icon* and *index*. The three types are listed here in decreasing order of conventionality. Symbols are highly conventional, iconic signs always involve some degree of conventionality, while indexical signs directly bring our attention to their objects.

An *icon* looks like what it signifies. An example of icons is a photograph.

An *index* has a causal and/or sequential relationship to its signified. For instance, indices are directly perceivable events that can act as a reference to events that are not directly perceivable. You may not see a fire, but you do see the smoke and that indicates to you that a fire is burning. Words *this*, *that*, *these* and *those* are also examples of indices.

A *symbol* represents something in a completely arbitrary relationship. Symbols are subjective. Their relation to the signified object can be dictated either by social and cultural conventions or by habit. Words are a prime example of signs.

This shows that each of the terms - icon, index and symbol - has two interpretations: (1) it is as a physical object with some meaning, and (2) it is a conceptual (theoretical) structure. We are interested here in the second meaning of these terms.

Besides, each of these concepts - *icon*, *index* and *symbol* – has the same structure as the concept *sign*.

In contrast to Saussure and Peirce, Morris (1938) defines *sign* in a dynamic way relative to some interpreter. He writes that *S* is a sign (M.B. more exactly, the sign name) of an object or objects *D* for an interpreter *I* to the degree that *I* takes the account of *D* in virtue of the presence of *S*. Thus, the object *S* becomes a sign only if somebody (an interpreter) interprets *S* as a sign (M.B. the sign name). This gives us the following diagram.

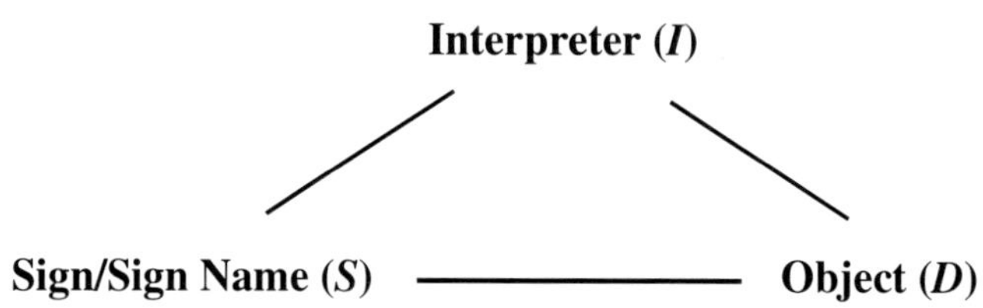

Figure 6.8. The *Dynamic Sign Triad* of Morris.

In addition to these basic components of sign, Morris also includes relations of the sign with other signs. This adds one more dimension, transforming the triad into a tetrahedron.

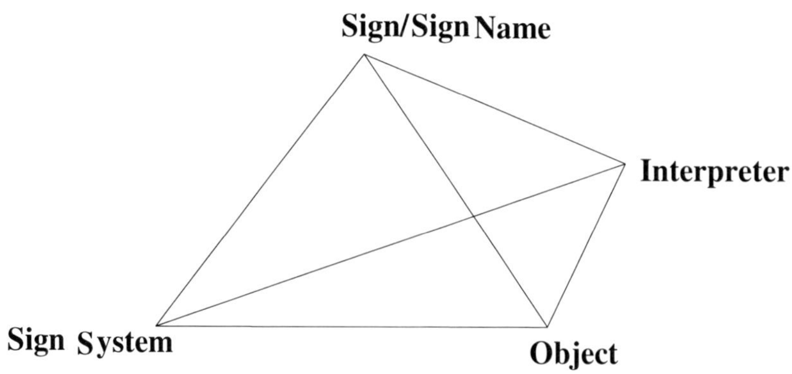

Figure 6.9. The *Dynamic Sign Tetrahedron* of Morris.

According to Morris, the *sign name* is what supports the triadic relation of the sign with other signs, with designated objects and with the subjects using the sign. These relations are represented by the corresponding fields of semiotics. *Syntactics* deals with sign relations to other signs in the Sign System, *semantics* is concerned with the study of designated objects and *pragmatics* is oriented at the subjects (interpreters) who use signs, whereas *semiotics* or *semiology* as a scientific discipline is occupied with the general study of *sign*.

Note also that Morris considers fuzzy relations because he takes in to account the *degree* that the interpreter *I* takes the account of the object(s) *D* in virtue of the presence of the sign name *S*.

This brings us to a variant of the Balanced Sign Triad of Peirce called the *semiotic triangle* (Figure 6.10) and has the following components (Nöth, 1990):

- *Sign vehicle* is the form of the sign.
- *Sense* is the sense made of the sign by the interpreter of the sign.
- *Referent* is what the sign "stands for".

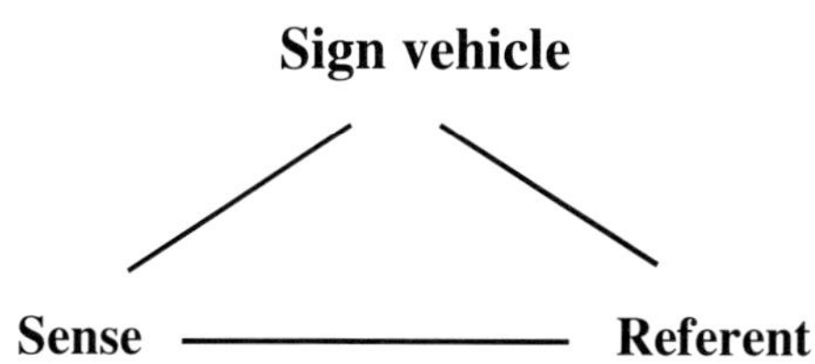

Figure 6.10. The Semiotic Triangle.

In this context, *sense* denotes the concept meaning for the interpreter of the sign.

Eco (1976) discerns three kinds of the sign vehicles, which are material signs:

- Signs for which there may be any number of tokens (replicas) of the same type, for example, a printed word.
- Signs whose tokens are different but similar, for example, a word which someone speaks or which is handwritten.
- Signs whose token is their type, or signs in which type and token are identical, for example, a unique original oil-painting or sculpture.

Another semiotic triangle (Figure 6.11) was suggested by Ogden and Richards (1923). Note that different versions of the Balanced Sign Triad of Peirce have been called by the name *semiotic triangle*.

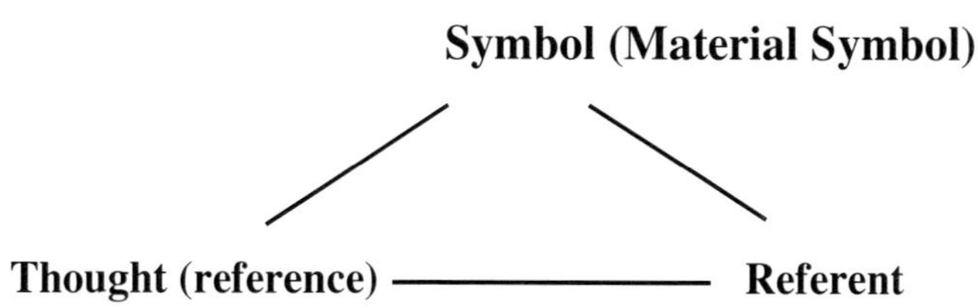

Figure 6.11. The *Semiotic Triangle* of Ogden and Richards.

Concepts and notions are important types of symbols. Indeed, a concept, or notion, is defined as a general idea derived from specific instances, i.e., a concept is a symbolic (usually, linguistic) representation of these instances and their unifying traits. This implies the fact that concepts/notions are signs/symbols of ideas have structures similar to the general structure of a symbol. Often a concept is used as a unit of knowledge and considered as a constituent of a proposition.

The first known model of a concept belongs to the noted logician Friedrich Ludwig Gottlob Frege (1848 –1925). According to him, concepts are ways of thinking of objects, properties and relations (Frege, 1891; 1892; 1892a). To explicate the structure of a concept, Frege suggested that in addition to having a *denotation*, names or descriptions of a concept also express a *sense*. Thus, the sense of a concept accounts for its cognitive significance — it is the way by which peopleconceives of the denotation of this concept. This gives us the following diagram.

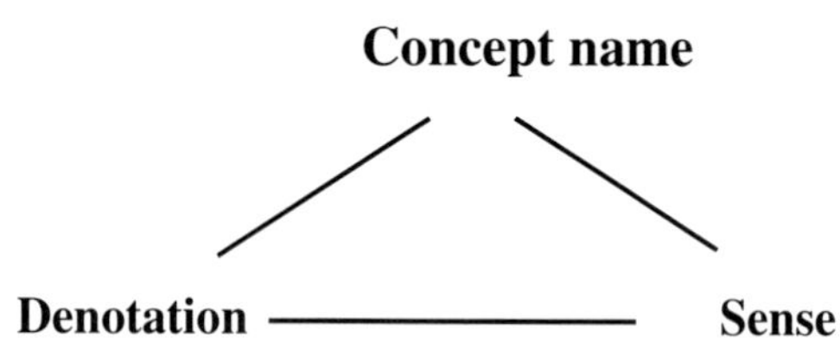

Figure 6.12. The *Concept Triangle* of Frege.

In addition, Frege defined *concept* also as a function, the value of which is always a truth-value (Frege, 1891).

Russell regarded concepts as constituents of propositions, while his model of *concept* is similar to the model of Frege. Considering concepts, Russell explained that some words (for example, proper names) *indicate* particular objects. This indication is represented by following structure.

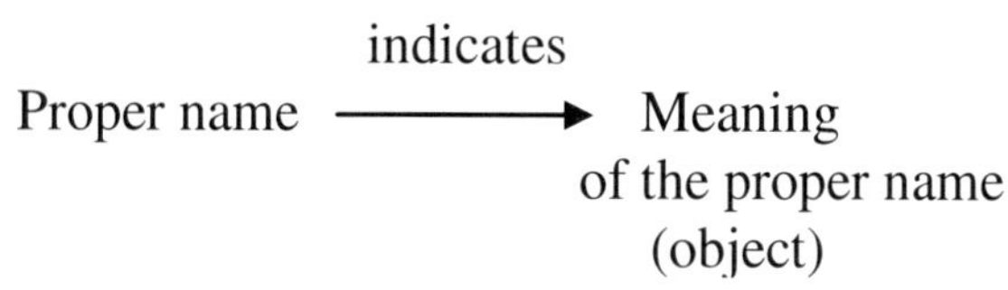

Thus, for Russell a concept has a name and two more constituents. On the one hand, concepts symbolize objects that are their exemplifications. Russell calls the relation between concepts and their particular exemplifications *denotation* (Russell, 1905). This relation is objective or, as Russell also says, logical. On the other hand, concepts have a part formed by *meanings* of corresponding linguistic expressions that *mean* objects denoted by the concept. This gives us the following structure of a concept.

It is interesting that explicating the structure of concepts, proper names and propositions in the theories of Russell and Frege, Tatievskaia (1999) demonstrates that they have the form of a fundamental triad described in Chapter 1. She portrays these structures as diagrams presented in Figure 6.14. This portrayal allows us to see that in the first diagram in Figure 6.14, *Sense* connects *Sentence* with *Reference* and in the second diagram, *Sense* connects

Proper Name with *Reference*, while the third diagram is a composition of two fundamental triads: in the first one, *Sense* connects *Concept-word* with *Reference* and in the second one, *Reference* connects *Sense* with *Referent.*

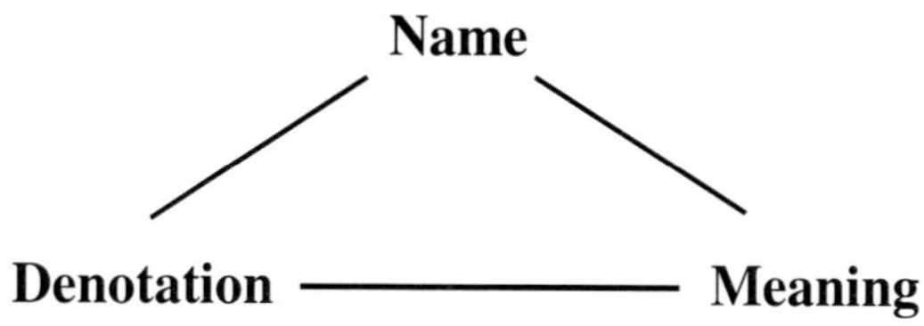

Figure 6.13. The *Concept Triangle* of Russell.

Cohen and Murphy (1984) consider four types of concept models: extensional, fuzzy set-theoretical, prototypical and semantic.

The *extensional* or *set-theoretical model* is based on set theory and represents a concept as a set of objects that belong to the concept, rather than as some form of mental representation. It is a classical model of concepts.

The *fuzzy extensional* or *fuzzy set-theoretical model* is based on fuzzy set theory and represents a concept as a fuzzy set of objects that to some extent belong to the concept.

The *semantic* or *knowledge-representation model* is based on componential semantics and usually represents a concept as a system of attributes in *feature-based models* and by conceptual networks in *network models*. These models of concepts are considered classical.

The *prototypical* or *prototype model* is based on *prototype theory* and represents a concept as a system of the role-values of a concept or a system of the most common attributes of the category. Such a system is considered as a prototype. Objects are denoted by the concept if they are sufficiently similar to this prototype (Rosch and Mervis, 1975).

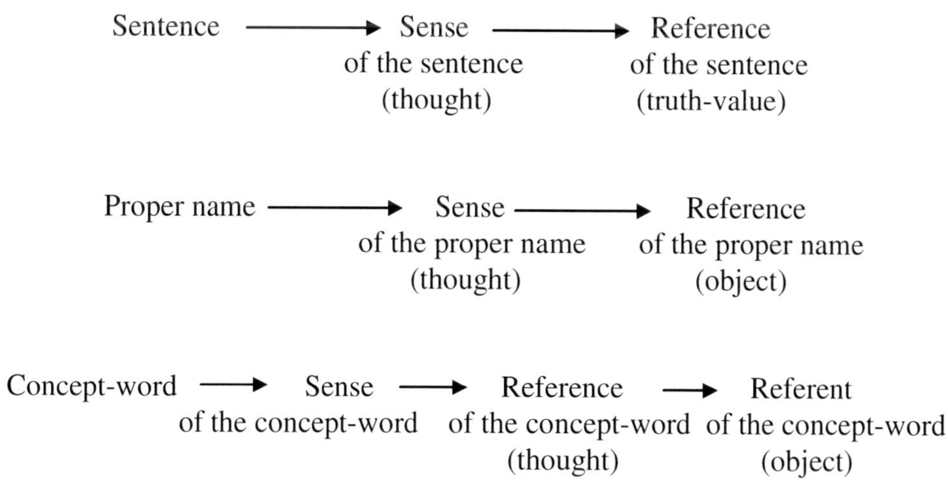

Figure 6.14. Conceptual structures of Frege.

These role-values of a concept are not assigned by definition because the concept does not specify necessary and sufficient criteria for membership. Objects that have the most typical role-values will be considered "good" prototypes, while objects that have atypical or unacceptable values will be considered "borderline" or "poor" prototypes. For instance, for the concept a *bird*, robin is a “good” prototype, while penguin is "poor" prototype.

Hampton (2011) writes that five broad classes of model have been proposed by different researchers.

The *classical model* assumes that concepts are clearly and entirely defined by a system of necessary and sufficient as the system attributes (Armstrong, *et al*, 1983; Osherson and Smith, 1981). Later the classical view was extended by dividing representation attributes into two groups: *defining features*, which form the core definition of the concept extension, and *characteristic features*, which are true only for typical category members and which may form the basis of a recognition procedure for quick categorization.

The second model is the *prototype model*, which is considered above.

The third one called the *exemplar model* is similar to the prototype model. In it, the concept is based not on one prototype, but on a number of different exemplar representations (Medin and Shoben, 1988).

The fourth model is the *theory-based model* (Murphy and Medin, 1985). It is rooted connections in the cognitive development tradition. Concepts are framed by the theoretical understanding of the world. While a prototype representation of a concept is a system of unconnected attributes, the theory-based representation has the form of a structured frame or schema, comprising theoretical knowledge about the relations between these attributes, as well as their causal and explanatory links.

The fifth model is the *psychological essentialism* (Medin and Ortony, 1989). It is the synthesis and development of the classical and theory-based models, and attempts to align psychological models with the philosophical intuitions. The model argues for a classical "core" definition for concepts, but one in which the core definition may frequently contain empty "place holders", for example, for additional still unknown attributes. In such a way, these "place holders" allows adding additional knowledge to concept definitions.

Burgin and Gorsky (1991) introduced a new model called the *representational model* of a concept. Its surface structure is presented in Figure 6.15, which is also a fundamental triad.

Concept Name ⟶ Conceptual Representative

Figure 6.15. The representational model of a concept.

We see that the *surface structure* of the representational model is similar to the structure of the model of a concept. However, the component *Conceptual Representative* is divided into two parts - *Interpretant* and *Denotation* (*Denotat*), forming the *concept structure of the second level*.

In turn, each component - *Interpretant* and *Denotation/Denotat* is divided into two parts, forming the *concept structure of the third level*: *Interpretant* consists of *Sense/Intention* and *Meaning*, while *Denotation/Denotat* consists of *Extent* and *Prototype*.

Then we proceed to the *concept structure of the fourth level* when a distinction is made between *literal components of the third level - literal Sense/Intention, literal Meaning, literal Extent* and *literal Prototype* - and *metaphorical components of the third level - metaphorical Sense/Intention, metaphorical Meaning, metaphorical Extent* and *metaphorical Prototype*.

In addition, we come to the *concept structure of the fifth level* when a distinction is made between strict *components of the third level* - strict *Sense/Intention*, strict *Meaning*, strict

Extent and strict *Prototype* - and fuzzy *components of the third level* - fuzzy *Sense/Intention*, fuzzy *Meaning*, fuzzy *Extent* and fuzzy *Prototype*. Each level includes all previous levels, making the model more detailed and more exact. The *concept structure of the fifth level* is presented in Figure 6.16.

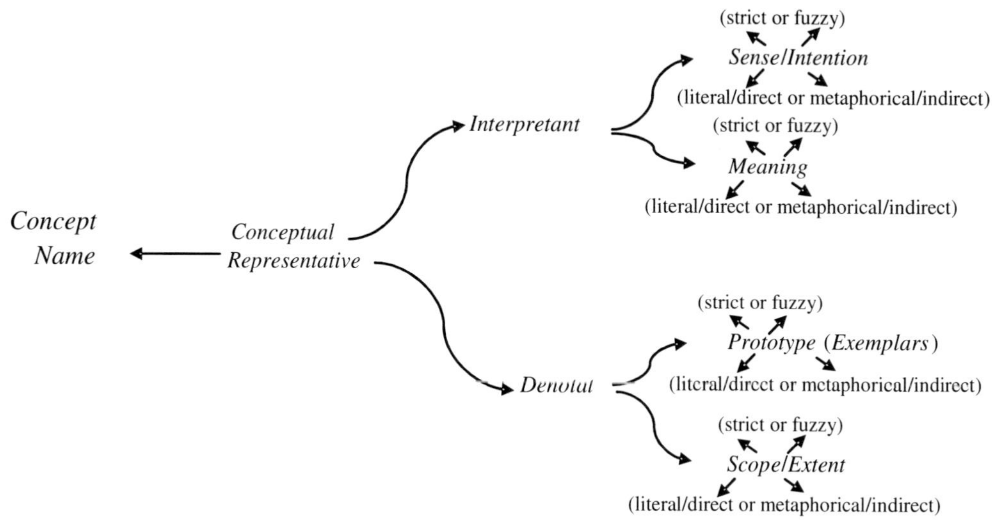

Figure 6.16. The concept structure of the fifth level in the representational model of a concept.

In this diagram,

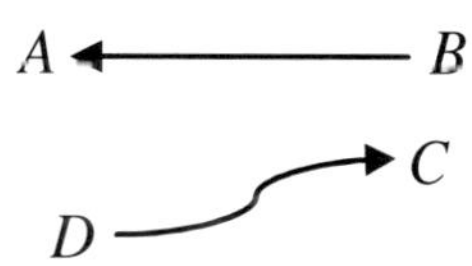

means "*A* is related to *B*."

means "*C* is a part (subclass) of *D*."

We see that the representational model of a concept comprises all other models of a concept.

Metaphorical Denotation (*metaphorical Denotat*) and *metaphorical Interpretant* form the *Concept Connotation* (*Connotat*).

With respect to its literal components – the literal meaning, literal extent, literal sense, and literal prototype, the concept plays the role of an *index*.

With respect to its metaphorical components – the metaphorical meaning, metaphorical extent, metaphorical sense, and metaphorical prototype, the concept plays the role of a *symbol*.

Besides, there are three forms of meaning:

- *Attributive meaning* has the form of a system of attributes.
- *Network* or *relational meaning* has the form of a conceptual network.
- *Operational meaning* has the form of a system of operations/procedures/algorithms.

It is necessary to remark that each component of the structure of a concept - the name, interpretant and denotat – is also an object and has a name, the name of this object. In addition, these objects also have an interpretant, i.e., sense and meaning, associated with their names. As a name is itself an object, it has a name and interpretant, i.e., sense and meaning, as any other object. In an intensional context, the names that occur denote the meaning or sense of the objects for the reader or listener. It means that each component of a concept can acquire the role of another component.

The same is true for the components of the structure of a sign - the name (sign vehicle), referent and interpretant/sense. Each of these components can acquire the role of another component. As a result, the structure of a sign, as well as the structure of a concept, has the property called *fractality*, which tells that the structure of the whole is repeated/reflected in the structure of its parts.

Chapter 7

THE STRUCTURE OF THE WORLD

The best intellectuals of the mankind have been always interested how the world is organized. For a long time, people did not have enough information about this world and used their imagination instead of knowledge, creating myths and legends. The higher level culture resulted in more elaborate myths and legends. An important component of these myths was the structure of the world where people lived.

For instance, in the classic Greek mythology, there was an elaborated cosmological structure with the following basic dyad, which is, as any other dyad, an implicit fundamental triad (see Chapter 1).

world ⟶ **underworld**

In turn, each component of this dyad had the triadic structure:

According to ancient Greeks, the world consisted of three parts: *Uranus* (sky), *Gaia* (earth), and *Pontus* (sea or ocean). This structure is presented in Figure 7.1. Zeus was the god of the sky and ruler of the Olympian gods. The brother of Zeus Poseidon was the god of the sea and protector of all waters.

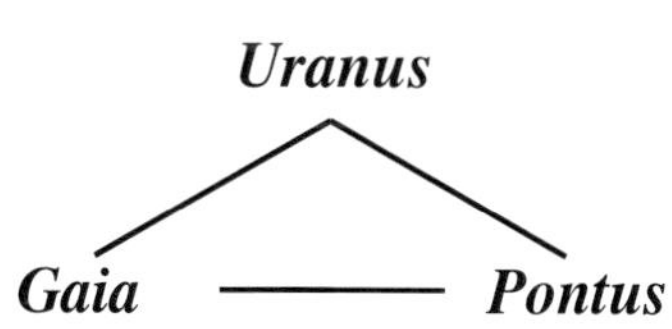

Figure 7.1. The world triad of the classic Greek mythology.

The underworld also consisted of three parts:

- The great pit of *Tartarus*, which was originally considered as the prison of the old Titan gods, later was assumed to be the dungeon home of damned souls.
- *The land of the dead* ruled by the god Hades, which is variously called the house or domain of the god Hades.
- *Elysium* or the *Elysian Fields* ruled by Rhadamanthys, where the virtuous dead and initiates in the ancient Mysteries were sent to dwell.

This structure is presented in Figure 7.2.

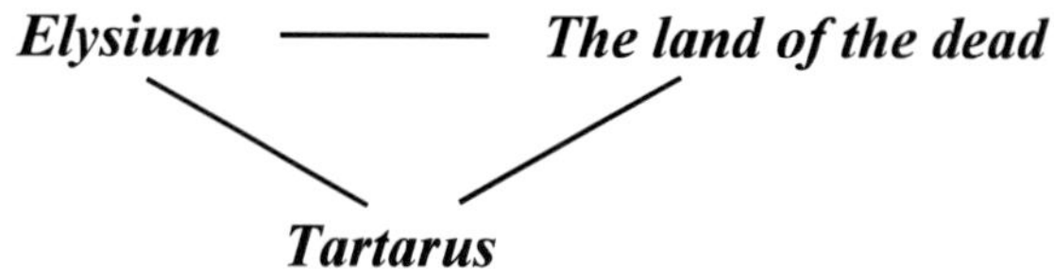

Figure 7.2. The underworld triad of the classic Greek mythology.

However, little by little people started to discern information obtained by reasoning and observation from information created by pure imagination. This was the beginning of philosophy.

It is usually acknowledged that Plato (429-347 BCE) was the first outstanding philosopher who constructed and discussed a systemic global structure of the world as a whole. In his dialogues, Plato mostly discusses two worlds: the World of Ideas/Forms, which is perfect unchangeable and eternal, and the Material World, which is not real being an imperfect reflection of the World of Ideas/Forms. The objects that are seen or in general sensed, according to Plato, are not real, but literally *mimic* the real Forms or Ideas. However, describing the World of Ideas/Forms, Plato cannot evade the crucial question asking how people get knowledge about this world. Answers inevitably involve the third world - mentality of an individual. Mentality is also necessary for comprehension of the Material World. In his dialogue *The Republic*, Plato has Socrates explain the literary analogy of a *divided line* to teach basic philosophical ideas about levels of existence and the corresponding structure of the individual mentality. Going from the lowest part to the highest, the individual mentality contains: *imagination*, *senses*, *intelligence* (as logical thinking) and (philosophical) *intuition*. In such a way, it is possible to conclude that in the worldview of Plato, the world as whole included the Mental World as its constituent. Thus, all three worlds amass to the *global world structure*, having the form of a triangular triad, the vertices of which are: the *material world*, the *mental world*, and the *world of ideas*, or *forms* (see Figure 7.3).

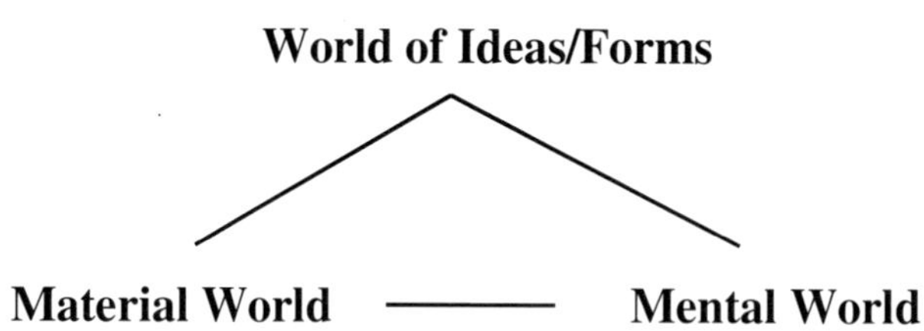

Figure 7.3. The Plato triad of the world.

As we know, Aristotle opposed Plato's approach rejecting existence of a separate World of Ideas. However, it is interesting that analyzing Aristotle's approach to understanding of the world, we can see a similar triad (see Figure 7.4), which is built of aspects of the material world instead of separate worlds. Indeed, Aristotle taught that each thing (physical object) is a composition of *substance* (*hyle*) and *form* (*eidos*) (see, (Aristotle, 1984)). Form comes into matter, but neither of them can exist independently in the physical world. Any form is based on a *prote hyle* (a "first matter"), which is the *hypokeimenon* (foundation) of the form. Forms, according to Aristotle, determine essence of things.

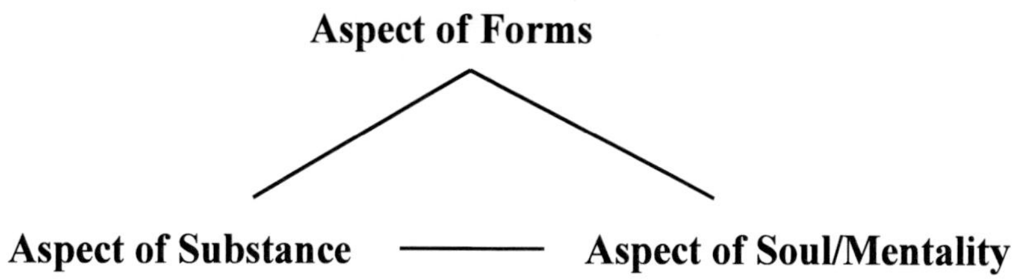

Figure 7.4. The Aristotle triad of the world.

Aristotle defines substance of a thing as "that out of which" this thing is made or as the "constituents" of this thing. As a result, substance is not necessarily physical. For instance, letters are the substance of words. Aristotle even calls the parts of a geometrical shape by the name *intelligible substance*.

Substance is only *potentiality* of things, which do not exist without form, which is *actuality* of things. According to Aristotle's theory of perception, the senses perceive an object by receiving its form. For instance, bronze is the matter of both a bronze statue and a bronze sphere. They have the same substance but different forms. When a bronze statue is transformed into a bronze sphere, the bronze as the substance remains constant throughout the change. While bronze is a bronze sphere's substance, *roundness* is its form. Bronze is potentially a bronze sphere. It becomes *actually* a bronze sphere when given roundness, which is an *actuality* of the bronze sphere. However, forms need not be shapes - *shape* is only a kind of forms. As the senses comprehend colors and flavors, forms include such properties as colors and flavors.

The triadic structure of Plato (Figure 7.3) has an interesting history because it has emerged in different outfits in works of outstanding thinkers. This history shows that to achieve a relevant understanding of the whole world is not a simple problem. According to physics, people live in the *physical* (*material*) *world* and this is the only reality that exists. In contrast to this opinion, subjective philosophies and religious teachings assert that only the mental world is real, while the physical world is just an appearance, a shadow without substance. For instance, Buddhism explains that the whole physical reality is a great illusion and the only reality is the spiritual world. In some forms of Buddhism (for example, *Yogacara*), it is assumed that all things are created by Mind.

In contrast to this, there are thinkers who believe that the true reality is an eternal unchanging world of forms or ideas. Plato was the first well-known proponent of this vision and many other have become his followers. Many scientists and mathematicians of our days share this opinion. Moreover, achievements of physics brought scientists to the edge of physical reality where extremely small physical objects, such as quarks or neutrons, and extremely large physical objects, such as our Universe as a whole, do not allow direct comprehensive observation. As a result, both extremities become more structures than material things. In this context, the outstanding physicist Max Born (1882–1970) admits that the notion of reality in the physical world had become, during the last century, rather problematic (Born, 1953).

At the same time, science has enough evidence to admit reality of the *mental world*. As states contemporary psychology, each individual has a specific inner world, which forms mentality of the individual and is based on the psyche. However, there is a controversy whether individual mentality is a product of the person's organism (body), or more exactly, of the brain, or the individual mentality transcends the body. In any case, these individual inner worlds form the lowest level of the mental world, which complements our physical world. On the next level, there are mental worlds of groups of people, communities and the whole society. It is possible to develop this hierarchy of mental worlds but it is done elsewhere demonstrating that the mental world is different from the physical world and constitutes an important part of our reality.

Moreover, our mentality influences the *physical world* and can change it. We can see how ideas change our planet, create many new things and destroy existing ones. Even physicists, who study the very foundation of the physical world, developed the, so-called, observer-created reality interpretation of quantum phenomena. The prominent physicist, John Archibald Wheeler (1911-2008), suggests that in such a way it is possible to change even the past. He stresses (Wheeler, 1977) that phenomena on the level of subatomic particles are unreal until observed. Existence of the *physical world* and *mental world* brings us to the dualistic model of reality.

However, the dualistic model is incomplete. This was prophesied in ancient Greece and proved by modern science. One of the great ideas of ancient Greek philosophy was the *world of ideas* (or *forms*), the existence of which was postulated by Plato. In spite of the attractive character of this idea, the majority of scientists and philosophers believe that the world of ideas does not exist, because nobody has been able to find any positive evidence in support of it. The crucial argument of physicists is that the main methods of verification in modern science are observations and experiments, and nobody has been able to find this world by means of observations and experiments.

Thus, having no evidence for and clear understanding of the World of Ideas, many philosophers and scientists, starting with Aristotle, argue that the World of Ideas causes many problems. Where is this world of ideas and how do we make contact with it? What is an idea in this sense? How is it possible for our mind to have an interaction with the Platonic realm so that our brain state is altered by that experience? Plato and his followers have not provided convincing answers to these questions. Thus, in spite of the attractive character of this idea, the majority of scientists and philosophers believe that the world of ideas does not exist, because nobody has been able to find any positive evidence in support of it. The crucial argument of physicists is that the main methods of verification in modern science are observations and experiments, and nobody has been able to find this world by means of observations and experiments. Nevertheless, there are outstanding modern thinkers, such as the philosopher Karl Raimund Popper (1902-1993), mathematician Kurt Gödel (1906-1978), and physicist Roger Penrose (1931-), who continue to believe that the world of ideas exists try to give a consistent interpretation of this world.

Thus, starting with ideas of Plato, Popper attempted to eliminate the incomprehensible World of Ideas/Forms (Popper, 1974; 1979) building his own triadic ontology of the world based on scientific ideas. It has two forms, which may be called the *general Popper Triad* and the *pure* Popper Triad.

In the pure form, Popper Triad consists of three parts/worlds:

- **World 1** consists of physical bodies, including microparticles, physical processes, physical energy and physical fields.
- **World 2** consists of thoughts, feelings, decisions, perceptions, observations, etc.

- **World 3** consists of abstract or intellectual products of the human mind, such as languages, tales, stories, contents of books and documents, movies and music, scientific conjectures and theories, mathematical constructions and engineering schemas, etc.

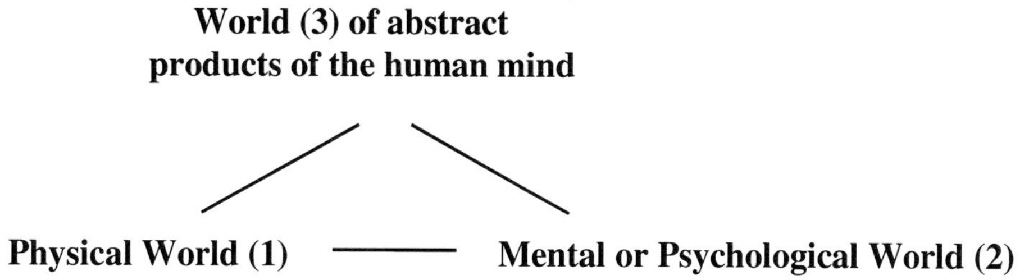

Figure 7.5. The Pure Popper Triad of the world.

To define these worlds, Popper (1976) writes

"If we call the world of "things" or of physical objects - the first world, and the world of subjective experiences (such as thought processes) the second world, we may call the world of statements in themselves the third world. …

It would be easy ... to regard the whole of world 3 as timeless, as Plato suggested of his world of Forms or Ideas.... I propose a different view - one which, I have found, is surprisingly fruitful. I regard world 3 as being essentially the product of the human mind.... More precisely, I regard the world 3 of problems, theories, and critical arguments as one of the results of the evolution of human language, and as acting back on this evolution."

However, such an abstract nature of World 3 caused criticism and many questions about it real existence. So Popper decided to extend this world, adding to it other products of the human mind, such as books and cars, paintings and sculptures, ships and planes, as well as other feats of engineering (Popper, 1978). As a result, World 3 acquired parts that also belonged to World 1. It is possible to try to separate World 3 and World 1, assuming that World 1 consists only of physical bodies that were not created by people. However, it is impossible to do this because now genetic engineers create various plants and animals, which after creation exist in nature in the same way as other plants and animals.

Thus, in the general form, Popper Triad consists of three parts/worlds:

World 1 consists of all kinds of physical bodies, including microparticles, physical processes, physical energy and physical fields.

World 2 consists of thoughts, feelings, decisions, perceptions, observations, etc.

World 3 consists of all products of the human mind, such as books, paintings, sculptures, planes and other feats of science and engineering

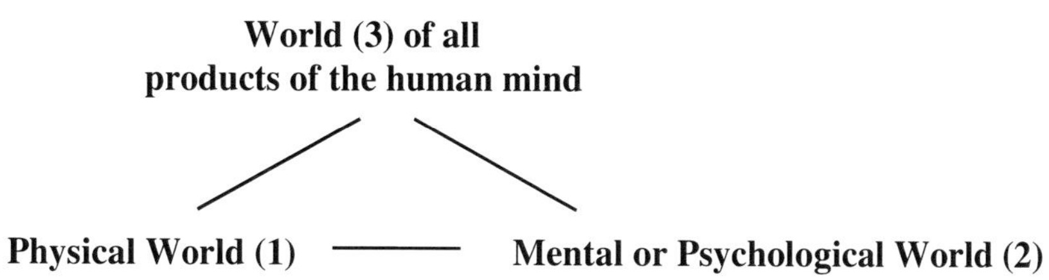

Figure 7.6. The General Popper Triad of the world.

On the one hand, it is much easier to understand Popper Triad than the Plato Triad because products of the human mind are much more tangible than abstract ideas, which exist nobody knows where. On the other hand, there are phenomena related to products of the human mind, such as shapes and other features of physical objects, which are not included in the World 3 of Popper.

Other authors refer World 3 in the triad of the world (in Figure 7.5) to signs in the sense of Charles Saunders Peirce (1839-1914), although they do not insists that it consists of objects that Peirce would classify as signs (see, for example, (Skagestad, 1993; Capuro and Hjorland, 2003)).

The prominent physicist Roger Penrose (1931) also considers three worlds and relationships between them (Penrose, 1994). One of the worlds is the mental world of the person. It is objective and real. Another world is the objective physical world. The third world is the world of mathematical truths. This gives us one more triad, which is called because the renowned mathematician Kurt Gödel had a similar conception of the global world structure.

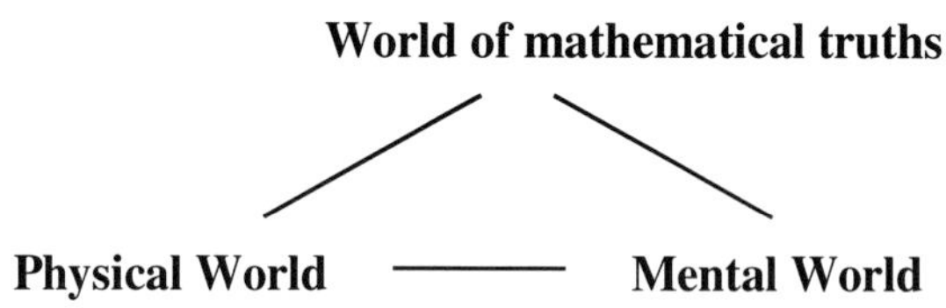

Figure 7.7. The Gödel-Penrose Triad of the world.

At the same time, the progress of science and mathematics brought forth the discovery of the *World of Structures* (Burgin, 1991b; 1996; 1997), allowing the researchers to solve the mystery of the Plato Forms or Ideas and to consistently represent World 3 from the Popper Triad. We can see that on the level of ideas, this world may be associated with the Platonic world of ideas or forms in the same way as atoms of modern physics may be related to the atoms of Democritus. At the same time, in contrast to Plato, science has been able to prove existence of the world of structures, demonstrating by means of observations and experiments, that this world constitutes the structural level of the world as the whole. Each system, phenomenon, or process either in nature, technology or society has some structure. These structures exist like material things, such as tables, chairs, or buildings do, and form the *structural level* of the world. When it is necessary to learn or to create some system or process, it is done, as a rule, by means of knowledge of the corresponding structure. As we have explained, structures determine the essence of things.

Thus, developing the tradition of Plato and Popper, Burgin (1996; 1997; 2010; 2011) introduced and studied the *Existential Triad* of the world as a whole. It is presented in Figure 7.8, giving a consistent understanding and scientifically grounded interpretation of Plato ideas.

Here the Physical (material) World is interpreted as the physical reality studied by natural sciences, the Mental World encompasses different levels of mentality, and the World of Structures consists of various forms and types of structures.

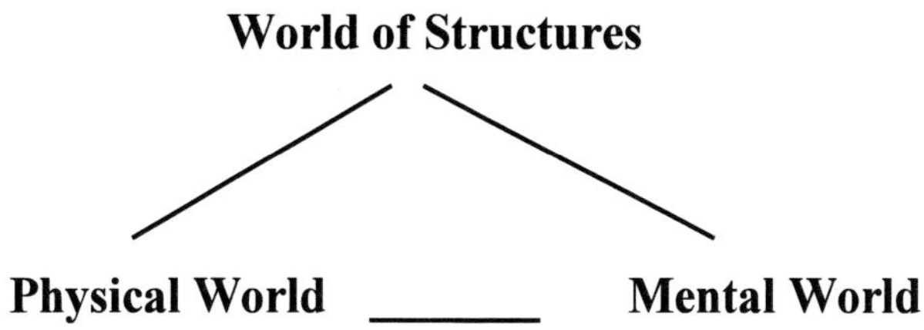

Figure 7.8. The Existential Triad of the World.

The Plato Triad looks very similar to the Existential Triad as there is a direct correspondence between vertices in these triads:

Material World ↔ Physical World
Mental World ↔ Mental World
World of Ideas/Forms ↔ World of Structures

So, it is important to understand what essential progress has been made from the time of Plato to our time and why these triads are similar and at the same time, fundamentally different.

Comparing the Material World of Plato and the Physical World from the Existential Triad, we see that on the ontological level, they are the same. However, on the epistemological (cognitive) level, they are basically different because due to the advancement of science, now people know much more about physical reality than even the best minds knew at the time of Plato. It means that the known physical world from the Existential Triad is many times larger than the known material world from the Plato Triad. In addition, knowledge of the physical reality has become more exact and comprehensive.

As in the case of the Physical World, contemporary knowledge about the Mental World exceeds what was known by Plato and its contemporaries. At the time of Plato, the Mental World included only individual mentality. Now science extended this picture and studies the Mental World on three levels, all of which are included in the Existential Triad:

- The first level treats mentality of separate individuals and is the subject of psychological studies. As in the case of physical reality, now psychology knows much more about mentality/psyche of people than even the best minds knew at the time of Plato. It is necessary also to remark that in the same way as physics does not study the physical reality as a whole but explores different parts and aspects of it, psychology also separates and investigates different parts and aspects of the mental reality, such as intelligence, emotions, or unconscious.
- The second level deals with group mentality of separate individuals and is the subject of social psychology. In particular, this level includes social conscience, which incorporates collective memory (Durkheim, 1984), collective intelligence (Brown and Lauder, 2000; Nguen, 2008a) and is projected on the collective unconscious in the sense Jung (see (Jung, 1969)) by the process of internalization. Some aspects of this level are studied by social psychology, which bridges sociology and conventional psychology.

- The third level encompasses society as a whole. Some aspects of this level are also studied by social psychology.

Besides, the Mental World from the Existential Triad comprises higher (than the third) levels of mentality although they are not yet studied by science (Burgin, 1997; 2010). It is possible to relate higher levels of the Mental World to the spiritual worlds described in many religious esoteric teachings.

....

↑

Social mentality

↑

Group mentality

↑

Individual mentality

Figure 7.9. The hierarchy of the Mental World (the *Basic Mental Hierarchy*).

The World of Ideas/Forms directly corresponds to the World of Structures because ideas or forms might be associated with structures. Indeed, on the level of ideas, it is possible to link ideas or forms to structures in the same way as atoms of Democritus and Leucippus may be related to atoms of modern physics. Only recently, modern science came to a new understanding of Plato ideas, representing the global world structure as the *Existential Triad* of the world, in which the World of Structures is much more understandable, exact and explored than the World of Ideas/Forms. When Plato and other adherents of the World of Ideas/Forms were asked what an idea or form was, they did not have a satisfactory answer. In contrast to this, many researchers have been analyzing and developing the concept of a structure (Ore, 1935; 1936; Bourbaki, 1948; 1957; 1960; Bucur and Deleanu, 1968; Corry, 1996; Burgin, 1997; Landry, 1999). It is possible to find the most thorough analysis and the most advanced concept of a structure in (Burgin, 2010). As a result, in contrast to Plato, science has been able to elaborate a sufficiently exact definition of a structure and to prove existence of the world of structures, demonstrating by means of observations and experiments, that this world constitutes the structural level of the world as the whole. Each system, phenomenon or process either in nature, technology or society has some structure. These structures exist like material things, such as tables, chairs, or buildings do, and form the *structural level* of the world. When it is necessary to learn or to create a system or to start a process, it is done, as a rule, by means of knowledge of the corresponding structure. Structures mold things in their being and comprehension.

If we say that structures exist only embodied in material things, then we have to admit that material things exist only in a structured form, i.e., matter (physical entities) cannot exist independently of structures. For instance, atomic structure influences how the atoms are bonded together, which in turn helps one to categorize materials as metals, ceramics, and polymers and permits us to draw some general conclusions concerning the mechanical properties and physical behavior of these three classes of materials. Even chaos has its structure and not a single one.

It is also necessary to remark that the World of structures includes Popper's World 3 from the Pure Popper Triad of the world because knowledge, as well as the intellectual contents of books, documents, scientific theories, etc., is a kind of structures that are represented in people's mentality (Burgin, 2010; 2011). This shows that the existential triad of the world is a further development of triads of Plato and Popper.

The three worlds from the Existential Triad are not separate realities: they interact and intersect. Individual mentality is based on the brain, which is a material thing, while in the opinion of many physicists mentality influences physical world (see, for example, (Herbert, 1987)). At the same time, our knowledge of the physical world to a great extent depends on interaction between mental and material worlds (see, for example, (von Bayer, 2004)).

In addition, there is a projection of the Mental World into the Physical World in the form of creations of human mentality (creativity), such as books, movies, magazines, newspapers, cars, planes, computers and computer networks. This projection determines the Extended Mental World, which consists of the Mental World and its projection. The Extended Mental World correlates with the World 3 from the General Popper Triad of the world.

Structural and material worlds are even more intertwined. Actually no material thing exists without a structure. Even chaos has its chaotic structure. Structures make things what they are. For instance, it is possible to make a table from different material: wood, plastics, iron, aluminum, etc. What all these tables have in common is not their material; it is specific peculiarities of their structure. In a similar way, as Poincaré (1908) wrote, space is in reality amorphous, and it is only the things that are in it that give it a structure (form). As some physicists argue, physics studies not physical systems as they are but structures of these systems, or physical structures. In some sciences, such as chemistry, and areas of practical activity, such as engineering, structures play the leading role. For instance, the spatial structure of atoms, chemical elements, and molecules determines many properties of these chemical systems.

Contemporary physics treats the physical world as a net of interacting components (systems), where there is no physical meaning to the state of an isolated object. A physical system (or, more precisely, its contingent state) is represented by the net of relations it entertains with the surrounding objects. As a result, the physical structure of the world is identified with such a global net of system relationships.

As North (2009) writes, physics is supposed to be telling us about the nature of the world, while physical theories are formulated in a mathematical language, using mathematical structures, which implies that mathematics is somehow telling us about the physical make-up of the world.

Plato postulated independent existence of his world of ideas, and it is possible to consider structures as scientific counterparts of Plato ideas. Thus, it is natural to ask the question whether structures exist without matter. Here we are not going into detailed consideration of this fundamental problem. What is important for the topic of this book is that as a coin has two sides, material things always have two aspects – *substance* and *structure*.

It is necessary to remark that it is possible to see existential triads in every person and in each computer. An individual has the physical component – her or his body (the physical component), the mental component studied by psychologists, and the structural component, which comprises all structures and systems from the other two components. A computer has the physical component - its hardware, the mental component, which consists of everything that is in computer memory, for example, software, and a structural component, which comprises all structures and systems from the other two components. It is also demonstrated that any natural language as a system has the structure of the existential triad (Burgin, 1997;

Burgin and Milov, 1998). The reason is that natural languages emerged (were created) as systems for reflecting and describing the whole world.

Thus, the three worlds from the Existential Triad are sides or basic aspects of the whole world. However, there is an interesting problem whether all three worlds can exist only in connection with one another or some (all) of them have an independent being. Contemporary science tells us that, at least, two of these worlds, material and mental, cannot exist without structures. Whether the Structural World has an independent being is the great problem of Plato and Aristotle. Science also does not give reliable evidence for mentality that is not based on physical systems in the same way as, for example, the mind is based on the brain.

There are also other approaches to the global Structure of the World. Many of them have the form of a triangular triad. For instance, Werner von Heisenberg (1901-1976) builds his world stratification based on relations between the states of things and knowledge processes. He writes (Heisenberg, 1942):

"We understand by "regions of reality" … an ensemble of nomological (i.e., representing laws of reality/nature, M.B.) connections. These regions are generated by groups of relations. They overlap, adjust, cross, always respecting the principle of non-contradiction. … It is clear that the ordering of the regions has to substitute the gross division of world into a subjective reality and an objective one and to stretch itself between these poles of subject and object in such a manner that at its inferior limit are the regions where we can completely objectify. In continuation, one has to join regions where the states of things could not be completely separated from the knowledge process during which we are identifying them. Finally, on the top, have to be the levels of Reality where the states of things are created only in connection with the knowledge process".

It is possible to represent Heisenberg's structure of the world. It has two forms of representation:

- as a linear/stratified triad (Figure 7.10.a), which a fundamental triad in the sense of (Burgin, 1998; 2011) where the middle element of the triad connects the first element with the second;
- as a triangular triad (Figure 7.10.b).

The linear form reflects a global stratification of the world into three layers, which form levels of reality. The triangular form demonstrates that all three components are closely connected.

The Heisenberg Triad is closely related to the Plato Triad, Popper Triad and Existential Triad. Its objective regions correspond to the material/physical world, while it is possible to relate knowledge-based regions to the world of structure (as knowledge is a structure (Burgin, 2010)) and to the mental world (as knowledge belongs to or is stored in mentality).

Another structure of the world was suggested by a prominent philosopher, biologist and sociologist Herbert Spenser (1820 –1903). It has two forms of representation:

- as a linear/stratified triad (Figure 7.11.a), which a fundamental triad in the sense of (Burgin, 1998; 2011) where the middle element of the triad connects the first element with the second;
- as a triangular triad (Figure 7.11.b).

Objective regions ⟶ **Regions with inseparable states of things and knowledge processes** ⟶ **Knowledge-based regions**

a)

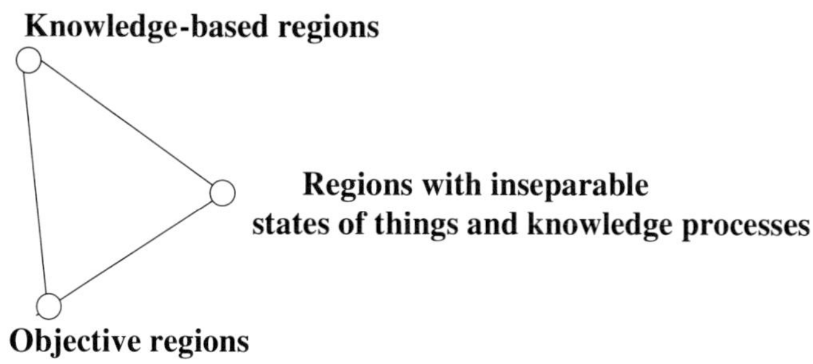

b)

Figure 7.10. The Heisenberg Triad of the World: a) the linear/stratified form; b) the triangular form.

The linear form reflects a global stratification of the world into three layers, which form levels of reality. The triangular form demonstrates that all three components are closely connected.

Inorganic world ⟶ **Organic world** ⟶ **Super-organic world**

a)

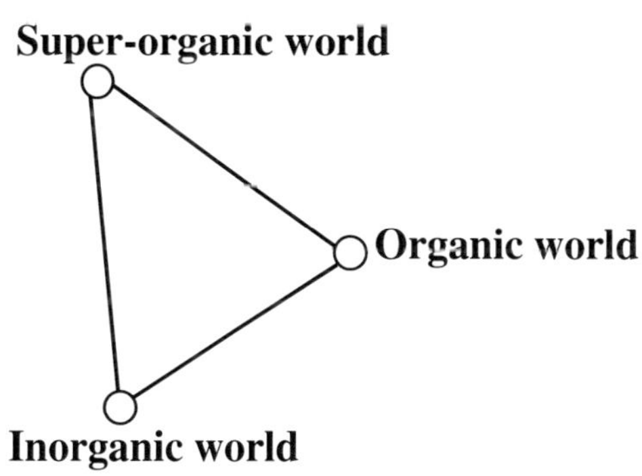

b)

Figure 7.11. The Spenser Triad of the World: a) the linear/stratified form; b) the triangular form.

The Spenser Triad is closely related to the Plato Triad and Popper Triad. Its organic and inorganic worlds belong to the material/physical world, while it is possible to interpret the super-organic world as both mentality and the world of ideas/knowledge.

The world stratification developed by Hartman (1940) and further specified by Poli (2006) in his theory of reality levels resembles the Spenser Triad of the World although its components are essentially different. Alike Heisenberg's and Spenser's Triads, the Hartman/Poli Triad of the World has two forms of representation:

- as a linear/stratified triad (Figure 7.12.a), which a fundamental triad in the sense of (Burgin, 1998; 2011) where the middle element of the triad connects the first element with the second;
- as a triangular triad (Figure 7.12.b).

The linear form reflects a global stratification of the world into three strata, which form levels of reality. The triangular form demonstrates that all three components are closely connected.

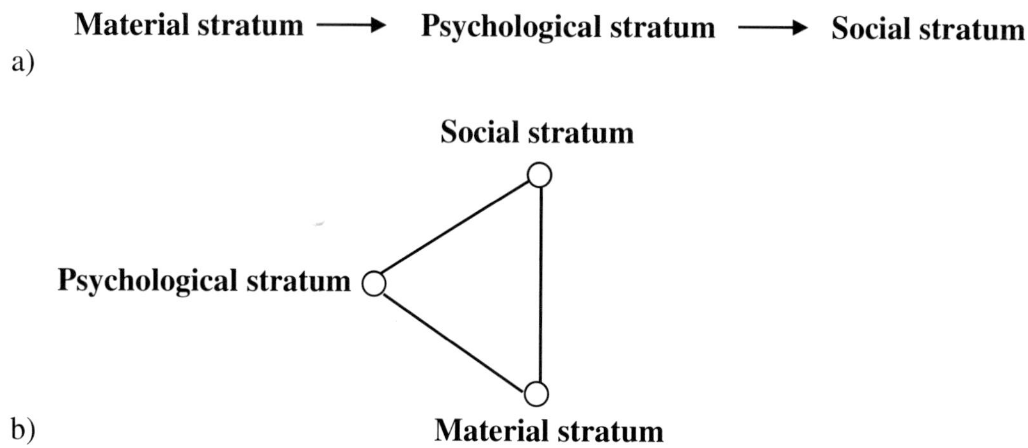

Figure 7.12. The Hartman/Poli Triad of the World: a) the linear/stratified form; b) the triangular form.

The Hartman/Poli Triad is closely related to the Plato Triad, Popper Triad and Existential Triad. The material stratum from the Hartman/Poli Triad is virtually the same as the material/physical world, the psychological stratum is similar to the mental/psychological world, while the social stratum may be related with the world of structures because social systems are more structures than physical entities. Thus, we can see the following system of correspondences:

Material stratum ↔ Physical World
Psychological stratum ↔ Mental World
Social stratum ↔ World of Structures

Morgan (1974; 1979) also built a triadic ontology of the world, postulating three parts/worlds. The structure of this ontology has two forms – linear and triangular. The linear form reflects a global stratification of the world into three layers/levels of reality. The triangular form demonstrates that all three components are closely connected.

The Morgan Triad is closely related to the Spenser Triad. In it, Matter corresponds to the organic and inorganic worlds from the Spenser Triad, Life is naturally related to the organic world and Mind is linked to the super-organic world.

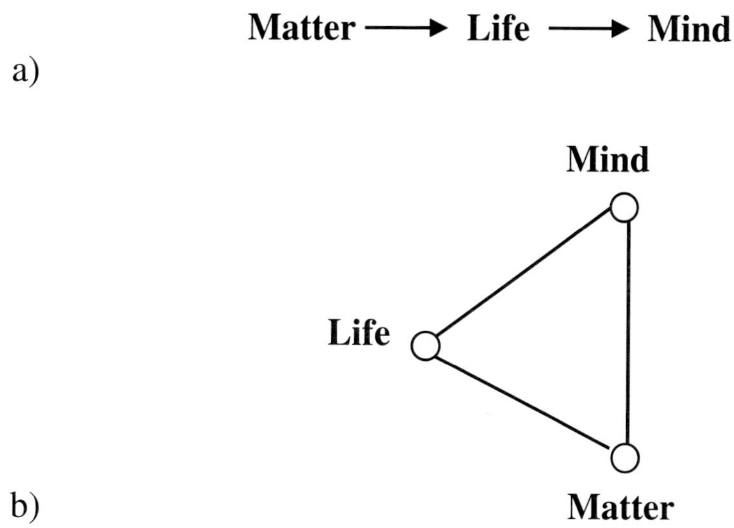

Figure 7.13. The Morgan Triad of the World: a) the linear/stratified form; b) the triangular form.

Goppold (1998) suggested the structure of the world in the form of the three existential world centers: SUB, OBJ, SEM. His approach is based on a grouping of all real world phenomena into the three existential world centers: *Subjective*, *Objective*, and *Intersubjective*. Such a structuring have often occurred under many different guises in the history of ideas

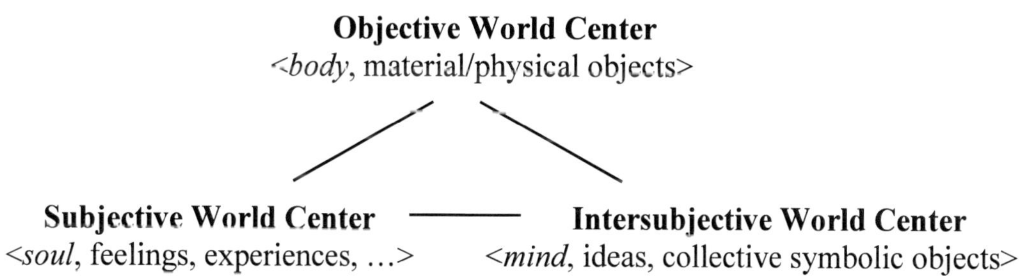

Figure 7.14. The Goppold Triad of the World in the form of the three existential world centers.

The Goppold Triad is isomorphic to the Plato Triad and to the Existential Triad, that is, there is a natural correspondence between components of these triads and between relations that connect these components. The Objective World Center directly corresponds to the material/physical world, the Subjective World Center is naturally related to the mental/psychological world and Intersubjective World Center is linked to the world of ideas in the Plato Triad or to the world of structures in the Existential Triad.

Triadic world structuration often allows further structuring of its components. For instance, the component Matter in the Morgan Triad (Figure 7.13) can be decomposed into four basic forms of matter: *solid state*, *liquid*, *gas* and *plasma*. At the same time, matter is stratified by the scale of reflection. Usually the following levels/scales are discerned:

Level/scale of filaments (supercluster complexes)
↑
Level/scale of superclusters
↑
Level/scale of galaxy clusters
↑
Level/scale of galaxies
↑
Level/scale of stars
↑
Level/scale of planets
↑
Level/scale of physical bodies with the size
similar to the size of mountains
↑
Level/scale of physical bodies with the size
similar to the size of people, for example, table, tree or plane
↑
Level/scale of physical objects
with the size microbes and viruses
↑
Level/scale of molecules
↑
Level/scale of atoms
↑
Level/scale of subatomic particles

The component Life in the Morgan Triad is stratified by biological classifications representing the following structure, which is a chain of fundamental triads (Burgin, 2011):

domains
(empires/super-kingdoms)
↑
kingdoms
↑
phyla/divisions
↑
classes
↑
orders
↑
families
↑
genera
↑
species
↑
organisms

Each level in this structure is also decomposed into several components. For instance, the top-level of this hierarchy consists of *Prokaryota* (or *Monera*), *Eukaryota* and the more recently discovered *Archaea* empires. The next level has six kingdoms: *Eubacteria*, *Archaebacteria*, *Protista*, *Fungi*, *Plantae* (*Plants*), and *Animalia* (*Animals*).

In a similar way, each world in the Existential Triad is a very complex system, which has an elaborated structure. Let us consider some components of these structures.

The basic structure of each world in the Existential Triad is based on the statics-dynamics duality, having two parts – substantial component and transformational component. They are connected by processes. This gives us three fundamental triads (see Chapter 1) presented in Figure 7.15.

Substance —**processes**→ **Energy**

Mental structures —**processes**→ **Mental Energy**

Structures —**processes**→ **Information in a strict sense**

Figure 7.15. The Basic Components of the World in the form of the Existential Prism.

It is interesting that defining information as "Information: that which determines form", MacKay (1969) also related information to the World of Structures because *forms* form a special kind of *structures* (Burgin, 2010).

The basic structure represented by triads in Figure 7.15 brings us to the Differentiated Structure of the World, which is called the *Existential Prism* (Burgin, 2011a). It is an amalgamation of the Existential triad and the energy/information reflection of it (see Figure 7.16).

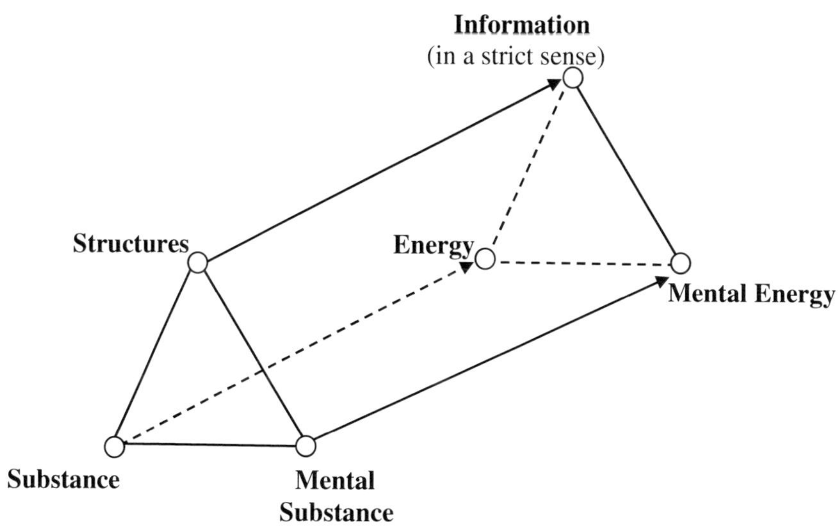

Figure 7.16. The Existential Prism.

Each of the three worlds from the Existential Triad has a hierarchical structure with several levels or strata. For instance, the *functional hierarchy* of the physical world goes from subatomic particles to atoms to molecules to bodies to cells to living beings and so on. As any other hierarchy, this one is represented by a chain of fundamental triads (Burgin, 2011) and is presented in Figure 7.17 Another hierarchy of the physical world called the *assembly hierarchy* is presented in Figure 7.18.

Level of groups of living beings
(social level)
↑
Level of separate living beings
↑
Level of cells
↑
Level of molecules
↑
Level of atoms
↑
Level of subatomic particles

Figure 7.17. The *Functional Hierarchy* of the Physical World.

Level of physical systems,
for example, human body, tree or plane
↑
Level of physical bodies,
for example, stones, tables or drops of water
↑
Level of molecules
↑
Level of atoms
↑
Level of subatomic particles

Figure 7.18. The *Assembly Hierarchy* of the Physical World (in nature).

There are also hierarchies of the Physical World related to scalable physics, which studies nature on different scales employing different theories.

In a similar way, it is possible to explicate hierarchies of the Mental World. One of them, the basic mental hierarchy, is described above (see Figure 7.8). Several hierarchics exist on each level of the basic mental hierarchy. One class of these hierarchies corresponds to the structure of personality on the individual, group and social levels. Personality hierarchies are described in Section 3.8.

Different authors also suggested hierarchical structures of the World. As Brenner (2008) writes, "the idea that reality is divided into levels that are more or less distinct and involve different degrees of complexity has been proposed, in various forms, since antiquity." Here we consider only contemporary authors.

Nicolai Hartmann (1882-1950) developed a multidimensional world structification, making a distinction between *levels of reality* and *levels of being*, suggesting three strata/levels of reality presented in Figure 7.12 and four strata/levels (*Seinsstufen*) of the sphere of real being (see Chapter 5): *matter* (material stratum), *life* (vital stratum), *consciousness* (psychical stratum), and *spirit* (spiritual stratum).

spirit
↑
consciousness
↑
life
↑
matter

Figure 7.19. Hartmann's stratification of real being.

Two other spheres of being, cognitive and logical, were also structured by Hartmann (1940). The cognitive sphere (*Erkenntnissein*) involves perception, intuition, knowledge, and comprehension (*Wahrnehmung, Anschauung, Erkenntnis, Wissen*).

intuition	→	knowledge
↑		↑
perception	→	comprehension

Figure 7.20. Hartmann's structurization of the cognitive sphere.

According to the tradition, logical sphere of being has three levels: concept/notion, judgment/statement, and inference/deduction.

inference/deduction
↑
judgment/statement
↑
concept/notion

Figure 7.21. Traditional stratification of the logical sphere.

Poli (2001; 2003; 2006) is developing a theory of levels of reality. In it, the term *level* is used as a general concept comprising two kinds of levels: *stratum*, which refers to the relams of being characterized by categorical diverse groups, and *layer*, which refers to the internal segmentation of strata. According to Poli, there are three strata of reality:

- the material stratum;
- the psychological stratum;
- the social stratum, which encompasses all phenomena of history, language, science, morals, in fact, the entire body of human knowledge and ideation.

Each stratum has its own principles, laws and ontological categories, and there are clear discontinuities between strata. In addition, hierarchical relations within strata are defined, separating levels of reality, which are governed by the same fundamental laws. Besides, *levels of organization*, which are different from levels of reality, are introduced. Several levels of organization or hierarchies can belong to one and the same level of reality.

In addition, Poli makes the further important distinction between *ontological levels of reality* and *epistemological levels of interpretation*. In his view, only some of the latter can be taken as levels of reality, namely those that are grounded on ontological categories. Levels of reality constrain the objects of the universe as to which types of causation and agency are admissible.

Nicolescu (2002) suggested that there are six major levels of reality:

- The *microphysical* or *quantum mechanical* level is characterized by non-thermodynamic (timeless) included middle states.
- The *macrophysical* level is characterized energetically by global entropy and gradual homogenization of its components.
- The *biological* level is characterized by local negentropy and the emergence of new forms (heterogenization).
- The *psychological-social* level contains sub-levels or nested levels to account for the ontological aspects of mental phenomena;
- The *cosmological* level contains interactions and emergent states of extreme complexity may be present.
- The level of "*Cyber-Space-Time*" is described as "both natural and artificial, potentially pre-existent but actualized at present, this "entity" at the interface between man and computer, with its source in the quantum world, involving a dimensionality higher than four, as well as a non-linear causality can be considered as a level separate from both the microphysical, the macrophysical and the (ordinary) psychological/conceptual."

To a scientist, the description of the world in terms of separate levels of reality seems natural and rigorous. An individual does not have direct contact with the world of quanta, or even cells but "visualizes" the structure of a proton, observes the fantastic variety of biological life, and is conscious of the existence of consciousness itself. This understanding supports a view of reality as constituted by levels that are in some essential respect separate. Different scientific laws apply to the phenomena at these different levels.

Brenner (2008) suggests another hierarchy that comprises six major levels of reality:

- The *microphysical* or *quantum mechanical level.*
- The *macrophysical level* is characterized energetically by global entropy and gradual homogenization of its components.
- The *biological level* is characterized by local negentropy and the emergence of new forms.
- The *human mental* or *psychological level.*
- The *human social level.*
- The *cosmological* (*universe* or *multiverse*) *level.*

Including information into the picture of the world allows us to explicate relations between the physical reality and mental world. Indeed, all people, including scientists, get their knowledge about the physical reality only based on information that comes (is obtained) from physical objects. Even theoretical and philosophical considerations become knowledge the physical reality only when they are correlated with information that comes (is obtained) from physical objects. As a result, information becomes an intermediary, a connecting link between physical and mental. Consequently, we have the following *Cartesian Triad*, which shows how the physical reality is reflected in mental world (see Figure 7.22). It is called the Cartesian Triad because Descartes was the main proponent of the dualistic approach to reality, separating it into the Material World and the Mental World.

Information

Physical (Body) ————→ **Mental** (Mind)

Figure 7.22. The Cartesian Triad.

At the same time, the lower levels of mentality (Burgin, 1997; 2010) are based on physical systems: on the brain in the case of the individual mentality and people in the case of the group and social mentality. This results in following Materialistic Triad. It is called the Materialistic Triad because it reduces mentality to the physical level of nature. It is necessary to remark that many people assume that mentality is a virtually independent from the physical reality.

Is based on

Mental (lower levels) ————→ **Physical**

Figure 7.23. The Materialistic Triad.

This shows that all three basic components of the world, the Physical World, Mental World and World of Structures, have a similar key structure, which is presented in Figure 7.24.

Substance ————→ **Information/Energy**

Figure 7.24. The Key Structure of the World Components.

There is such an interesting phenomenon as *fractality* when the whole is reflected in its parts and this peculiarity persists as we are taking smaller and smaller parts. By analysis, we can discover structural fractality of the Existential Triad. Namely, we find that the Existential Triad is reflected in many other basic structures, which are structures of the parts or domains of the world.

One of these basic structures is the structure of a sign represented by the *Balanced Sign Triad* of Peirce (see Figure 6.7). In it:

- the component *Sign Name* fits in the World of Structures as a syntactic system (note that a name may consists of many words and contain relations between them, see (Burgin, 2011));
- the component *Sign Object* (*Denotat*) corresponds (but does not necessarily belong) to the physical world;
- the component *Sign Meaning* (*Interpretant*) belongs to the mental world because mentality creates meaning as it is necessary to think in order to understand.

In many cases, the object is a material thing and as such, is a part of the physical world. However, object can be non-material and thus, it does not belong to the physical world in some cases. For instance, the word *joy* is the name of emotion, which is nonmaterial. The word *theorem* is the name of a mathematical statement with definite properties. A statement is also nonmaterial. Nevertheless, *object* as a component of sign plays the same role as thing, implying that the Balanced Sign Triad is structurally isomorphic to the Existential Triad. In other words, the Balanced Sign Triad is homologous, i.e., essentially similar, to the Existential Triad of the World.

The same is true for the *Dynamic Sign Triad* of Morris (see Figure 6.8):

- the component *Sign Name* corresponds to the structural world as a syntactic system;
- the component *Object* corresponds (but does not necessarily belongs) to the physical world;
- the component *Interpreter* corresponds to the mental world as the interpreter interprets sign in his/her mentality.

This shows that the Dynamic Sign Triad of Morris is also structurally isomorphic to the Existential Triad.

One more triad emerged in the *Frege–Church ontology*. We remind that being a part of metaphysics, *ontology* is the philosophical study of the nature of *being, existence* and *reality*, concentrating on the basic categories of being and their mutual relations.

The *Frege–Church ontology* was developed by Alonzo Church (1951) based on ideas of Gottlob Frege to resolve paradoxes with language utilization. According to this ontology everything is considered as being in three categories: *object* (*referent* or *denotation*), *name* and *concept* (*sense*). In particular, each *word* is a *name* and has a *name*, a *denotation*, and a *concept* associated with it.

One more basic structure is the structure of language in the sense of Saussure. As the main tool for people's communication, language has to represent the structure and peculiarities of the world where people live. This idea was expressed by Bertrand Russell (1903) and indeed, the structure of language mirrors the Existential Triad of the world.

As we know, on the level of texts, as well as on the level of words, any language has three components: syntax (morphology), semantics and pragmatics. They

On the level of words, these components have the following meaning.

(1) *Morphology* is the study of word formation from morphemes, which are the smallest meaningful elements of a language.

(2) *Semantics* describes meaning of words.

(3) *Pragmatics* reflects and investigates the impact of words on people.

These components form the structure presented in Figure 7.25.

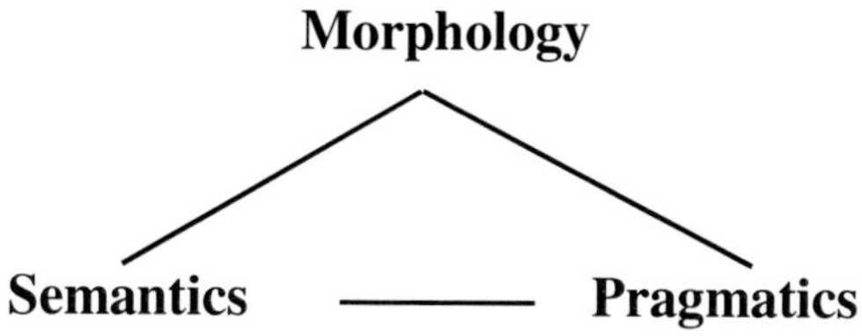

Figure 7.25. The *Lexical Structural Triad of Language.*

This triad is homologous, i.e., essentially similar, to the Existential Triad. Namely, morphology corresponds to (reflects) the World of Structures, pragmatics belongs to the Mental World, and semantics relates to the Physical World.

Note that here and later the correspondence between semantics and the Physical World does not mean that semantic systems are physical. However, semantics assigns meaning to syntactic structures in a similar way as substance provides a physical realization for structures of physical systems.

On the level of texts, language also has three components – syntax, semantics and pragmatics, which have the following meaning.

(1) The *syntax* of a language means the system of the rules that govern the structure of texts in this language. In such a way, syntax represents relations of linguistic entities to each other and determines its structure.

Here are two examples of such rules and relations from the syntax of English.

1. There are two basic parts to every sentence: the subject and the predicate.
2. The predicate (often in the form of a verb) goes after the subject (often in the form of a noun) in a sentence.

(2) The *semantics* of a language represents meaning of compound linguistic entities (words, sentences, texts). There are many meanings of the term *meaning*. Even in theory, there are different approaches to interpretation of this term and there is no generally accepted consensus. For instance, Ogden and Richards (1953) found 23 meanings of this term.

The *pragmatics* (also called *dectics*) reflects and investigates the impact (action) of texts on people who use language.

These components form the structure presented in Figure 7.26.

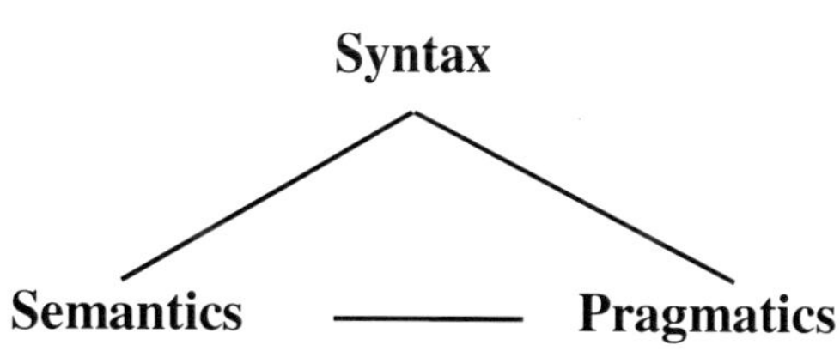

Figure 7.26. The *Main Structural Triad of Language.*

This triad is also homologous, i.e., essentially similar, to the Existential Triad (Burgin and Milov, 1998)). Namely, syntax corresponds to (reflects) the World of Structures, pragmatics belongs to the Mental World, and semantics relates to the Physical World.

All three components of the Main Structural Triad of language hinge on various relations. The semantics deals with relations between signs and objects that signs designate. The syntax studies and describes the relations between signs. The pragmatics treats relations between signs and people who use them.

We can also discover fractality in the Structural Triad of language.

There are three basic sorts of meaning:

Extentional or *denotational meaning* of a linguistic object (word, sentence or text) is the relation of this linguistic object to a non-linguistic object(s), which is traditionally called the *denotat* or *denotation*.

Intentional or *connotational* or *structural meaning* is the relation the signs have to mental signs that are conceived of as mental states or concepts. Intentional meaning presented by the relation between a given sign and the mental signs that serve in practical interpretations of the given sign is traditionally called *sign connotation*.

Referential or *compositional meaning* refers signs to other signs constructing the meaning of the sign from meanings of other signs.

As a consequence, we obtain three main types of semantics:

Denotational semantics refers linguistic objects (words, sentences, texts) to non-linguistic objects.

Intentional/Connotational semantics refers linguistic objects (words, sentences, texts) to other mental objects.

Referential/Compositional semantics refers linguistic objects (words, sentences, texts) to other linguistic objects.

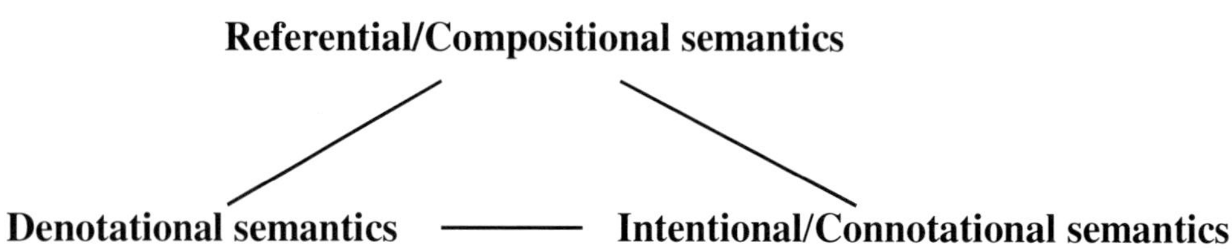

Figure 7.27. The *Structural Triad of Semantics*.

This triad is also homologous, i.e., essentially similar, to the Existential Triad. Namely, compositional semantics corresponds to the World of Structures, intentional semantics reflects the Mental World, and denotational semantics is associated to the Physical World.

A similar situation exists in syntax, which is also stratified according to the Main Structural Triad of language.

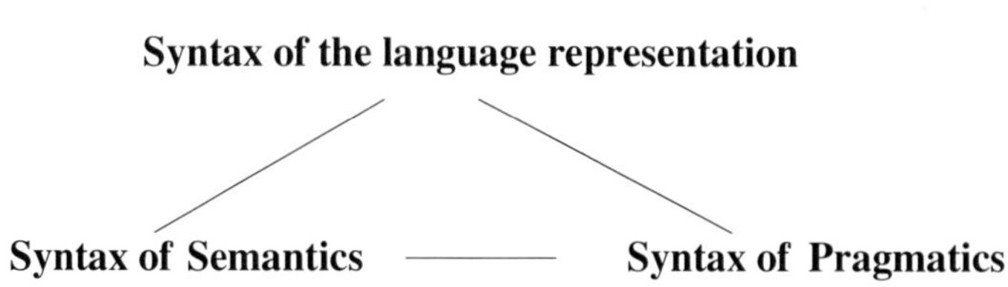

Figure 7.28. The *Structural Triad of Syntax*.

This triad is also homologous, i.e., essentially similar, to the Existential Triad.

The components of the Structural Triad of Syntax have the following functions:

The syntax of the language representation describes how syntactic rules are applied in construction of linguistic structures.

The syntax of pragmatics defines generation of pragmatic structures.

The syntax of semantics defines generation of semantic structures.

The Structural Triad of Semantics (Figure 7.27) induces a decomposition of the syntax of semantics generating the following triad:

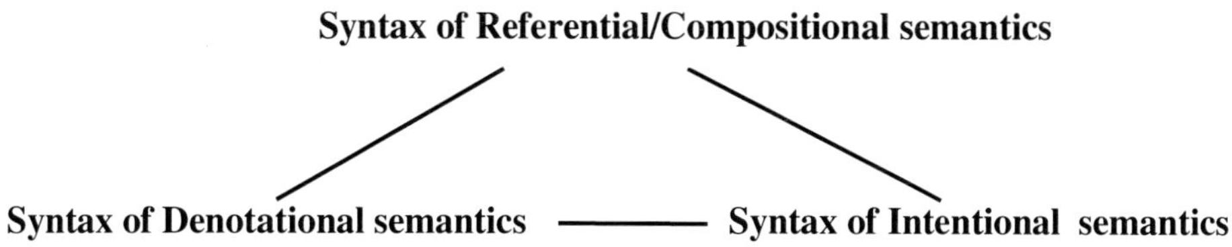

Figure 7.29. The *Structural Triad of Semantic Syntax.*

This triad is also homologous, i.e., essentially similar, to the Existential Triad.

There is also a decomposition of pragmatics induced by the Main Structural Triad of Language (Figure 7.30).

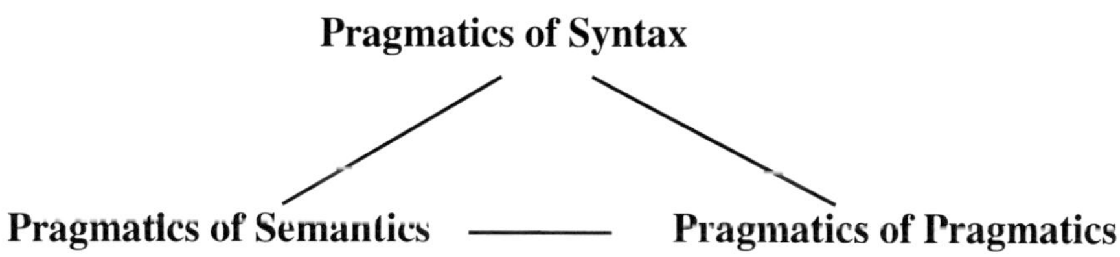

Figure 7.30. The *Structural Triad of Pragmatics.*

The components of the Structural Triad of Pragmatics have the following functions:

The *pragmatics of syntax* shows how to build texts in general and sentences in particular so that it would allow one to get the necessary influence.

The *pragmatics of semantics* reflects connections between meaning in general and personal meaning, showing how general meaning is transformed in the mentality of people.

The *pragmatics of pragmatics* is the higher pragmatics, which shows how direct impact of communication is transformed in the mentality of people during their life.

This triad is also homologous, i.e., essentially similar, to the Existential Triad. Indeed, the Syntax of language representation corresponds to the World of Structures, syntax of pragmatics reflects the Mental World, and syntax of semantics is associated to the Physical World.

Lacan (1977) applied the structural analysis to psychoanalysis and came to a reflection of the Existential Triad. Subconscious (in his opinion) is structured as a natural language. It is possible to describe his main conception by the triad:

Real - Imaginary – Symbolic

Real is chaos and *Imaginary* is an individual variation of the *symbolic order*, which is objective. In the context of the Existential Triad, the Lacan's Triad is interpreted in the following way:

- Real is related to the Physical/Material World.
- Imaginary is corresponds to the Mental World.
- Symbolic is fits into the World of Structures.

The Lacan's Triad has the structure of a fundamental triad because *Imaginary* connects *Real* with the *Symbolic*.

Lacan also suggested the thesis that an idea and existence are not identical: it is a language, which is a mediator between them. This gives us one more fundamental triad:

$$\text{idea} \rightarrow \text{language} \rightarrow \text{existence}$$

Tall (2004; 2004a) introduced three worlds of mathematics, also coming to a reflection of the Existential Triad in mathematics:

- *Conceptual-embodied mathematics world* grows out of our perceptions of the world and consists of our thinking about things that we perceive, sense and associate with mathematical concepts, including not only mental perceptions of real-world objects, but also internal conceptions that involve visuospatial imagery.
- *Proceptual-symbolic mathematics world* is the world of symbols used for calculation and manipulation in arithmetic, algebra, calculus and so on, enabling us to switch fluently from processes to do and concepts to think about.
- *Formal-axiomatic mathematics world* is based on *properties*, expressed in terms of formal definitions and axioms specifying mathematical structures (such as 'group', 'field', 'vector space', 'topological space' and so on).

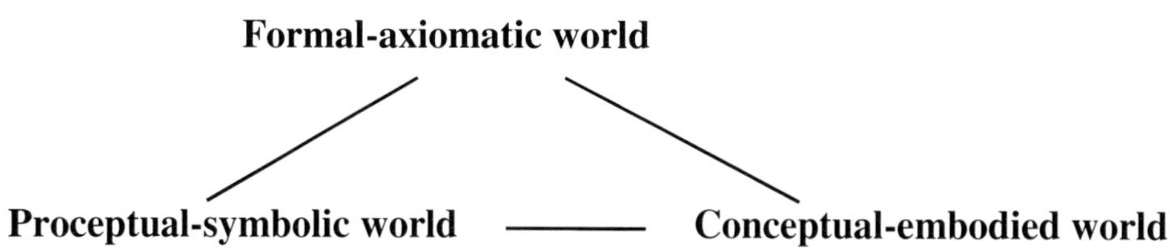

Figure 7.31. The Structural Triad of the World of Mathematics.

There is a direct correlation between these three worlds and the Existential Triad. Indeed, it is easy to see that thinking, perceptions and visuospatial imagery are components of the individual Mental World. Thus, the conceptual-embodied world naturally corresponds to the Mental World. Mathematical structures, as it is demonstrated in Chapter 2, belong to the World of Structures, i.e., the formal-axiomatic world is logically associated with the World of Structures. The proceptual-symbolic world consists of symbols used for calculation and manipulation, while the notion of *procept* builds initially on actions in the embodied world and the initial stages of counting and early arithmetic are largely embodied. It means that signs in the proceptual-symbolic world are material entities (see Chapter 6). Thus, the proceptual-symbolic world naturally corresponds to the Physical World. This concludes our demonstration that the World of Mathematics is globally organized as the Existential Triad.

It is interesting that the Existential Triad is also reflected in the three basic functions of the brain:

- Reasoning as symbolic information processing and information processing of images.
- Emotions and feelings as affective states of the brain.
- Will and instinct as vital forces of the psyche.

These functions form one more triangular triad.

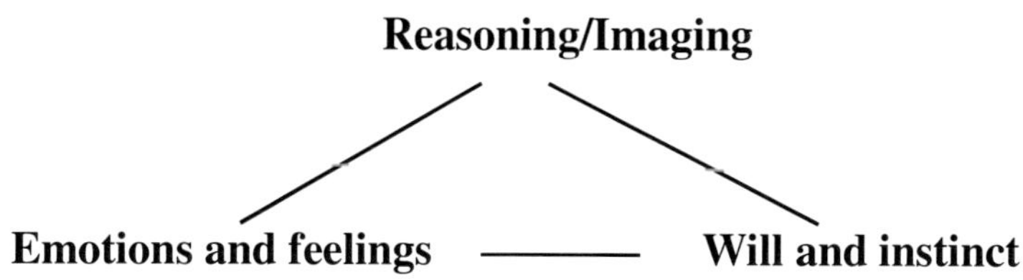

Figure 7.32. The Brain Functional Triad.

All these functions belong to the Mental World. However, the Brain Functional Triad is isomorphic (homologues) to the Existential Triad, i.e., there is a natural similarity (correspondence) between these triads. Reasoning/Imaging corresponds to the World of Structures as it consists of various operations with structures. Emotions and feelings correspond to the Physical World as (affective) states of the organism and psyche (Burgin, 2010). Will and instinct correspond to the Mental World as they are vital forces of mentality (Burgin, 2010).

As it is demonstrated in (Burgin, 2010), the Brain Functional Triad has its computer counterpart, which is presented in Figure 7.33 and is also homologues to the Existential Triad of the World.

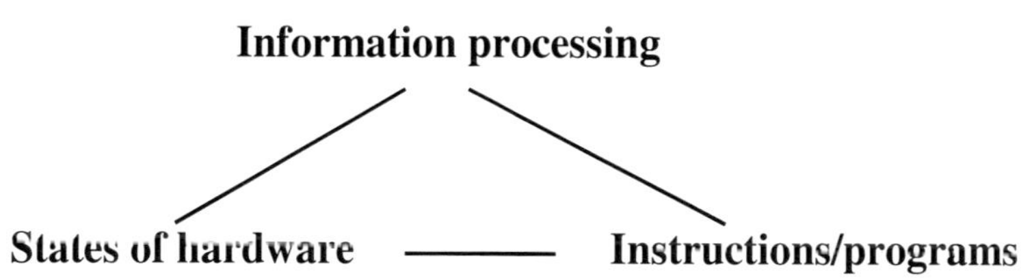

Figure 7.33. The Computer Functional Triad.

Different triads related to the living cell of an organism are also isomorphic to the Existential Triad of the World as it is demonstrated in (Ji, 2001; 2002; 2012).

All these examples show how the structure of the world as a whole is reflected in different parts of the world, explicating fractality of this structure because the structure of the whole is mirrored in the structure of its parts.

The Existential Triad of the World is also reflected in human culture, which has the *material component*, *mental/spiritual constituent* and *structural ingredient*. At the same time, culture as a whole consists of a variety of local cultures, bounded in time and localized in place on the Earth. Many local cultures existed before and exist now. For instance, we are speaking about ancient Greek and modern American cultures, about medieval European and ancient Chinese cultures, culture of native Americans and culture of primitive tribes, and so on and so forth. The huge variety of cultures on the Earth is separated into several groups to form the global structure of human cultures.

The traditional approach is based on the opposition *West-East*, which is connected to the two opposite directions on the globe, the East and the West. This scheme provides a good framework for comparison of European and some Asian cultures but causes many problems due to its incompleteness. For instance, is Islamic culture a Western or Eastern culture? With respect to Europe, it is an Eastern culture, but not to such extent inherently Eastern as Chinese culture. At the same time, with respect to Japan, Islamic culture is a Western culture, but not really Western in comparison with the French culture. Another problem with this typology is that not all cultures have their place in it. As an example, we can take an African culture, which is evidently neither a Western culture nor an Eastern culture.

Consequently, we come to a conclusion that to reflect adequately the existing diversity of cultures, it is necessary to elaborate a more developed global classification than that consisting only of two groups, East and West. As a base for a new typology, it is natural to take the traditional structure of the directions on the Earth: *North, South, East* and *West*.

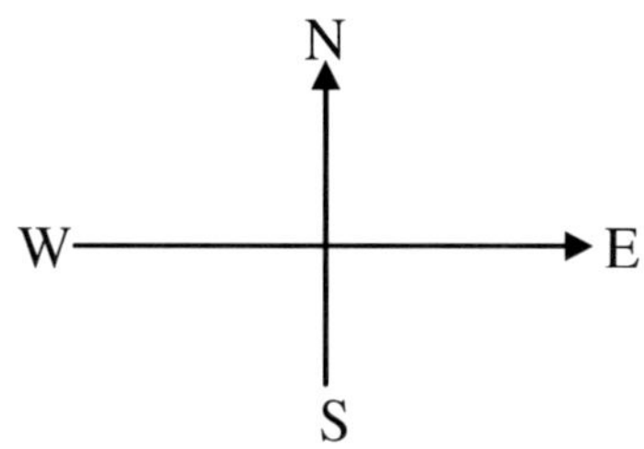

In such a way, we obtain four groups of: *Northern, Southern, Eastern* and *Western* cultures.

To make the classification complete, we need one more point that is called *Center*. The *Central Culture* includes Jewish, Islamic, and, to some extent, Persian cultures. European and modern North American cultures represent *Northern culture* although traditionally it is called Western culture. Cultures of India, China, and Japan form the nucleus of *Eastern culture*. *Southern culture* is mainly the culture of Central and South Africa. In this context, *Western culture* is the culture of Native Americans, the main body of which under the influence of Spanish/Portuguese culture has been transformed into the culture of Latin America.

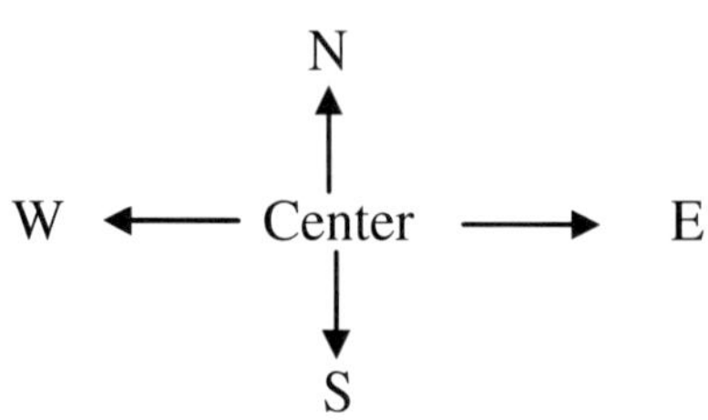

Figure 7.34. The structure of human culture.

However, here, we are not going to analyze the whole system of human cultures in all its complexity. Our aim is to expose briefly oppositions and similarities that exist between Eastern and Western (Northern with respect to the new classification) cultures and to show what conclusions it implies.

At first, it is necessary to make two remarks. First, traditionally researchers in the field are concentrated on the existing dissimilarities. We are treating the problem from a different perspective emphasizing similarities as the base for cultural interaction. Second, in modern society, traditional distinctions between various cultures are becoming vaguer and vaguer due to the rapid increase of interactions between different countries and people. However, even now the essential traits of different cultures determine a lot in thinking, attitude, and comprehension of people who belong to a definite culture. To understand each other, people have to know about these traits.

The most important inherent similarity between Eastern and Western (Northern) cultures is that *they are both of the intellectual type*. It means that the main engine of these cultures is the rational human intelligence. Consequently, they are oriented at knowledge. It is necessary to remark that such an orientation of these cultures has been a result of their achieving a sufficiently high civilization level.

However, intellectual orientation of these cultures is different. In Western (Northern) intellect and intelligence, the left hemisphere of the brain is dominant. This hemisphere, according to modern neurological theories, processes symbolic information. It is connected to the linguistic representation of information, step-by-step reasoning and mostly sequential algorithms of information processing.

In Eastern intellect and intelligence, the right hemisphere of the brain is dominant. This hemisphere, according to modern neurological theories, processes images (or, as it is called in psychology, *gestalts*). It is connected to the holistic representation of information and mostly parallel algorithms of information processing. Some cultural practices make stress on these distinctions. For example, training in Zen-Buddhism includes such an element when it is demanded to think not by the brain but by the stomach.

It is interesting that all other dissimilarities between these two types of cultures may be deduced from this principal distinction. Let us do this, taking into account the following aspects: style of life; typical attitude and comprehension; and prevalent thinking and behavior. Note that the given description of distinctions and commonalities between Western and Eastern cultures is not correct for all individuals that belong to these cultures. There are many exceptions. Besides, contemporary globalization diminishes differences, at least, on the surface of the modern society. The given description reflect only basic aspects and modes of these cultures.

The left hemisphere has dominated the brain of the Westerner, implying abstract, and therefore, exact thinking and behavior. Western mentality is oriented at material things and is based on individualistic ideas and views.

The right hemisphere dominated the brain of the Easterner, implying concrete, and therefore, fuzzy thinking and behavior. Eastern mentality is oriented at the inner world of an individual and is based on collectivism as an inherent principle. In Eastern cultures, as a rule, individuals perceive themselves as members of a group, prioritizing the group rather than themselves.

As a corollary, we find that Eastern cultures are more inertial, adaptive, conservative, and stable if sufficiently big. At the same time, we can see that Western cultures are more flexible, innovative, adapting, and continuously changing.

The discrete information processing of the Westerner implies systemic, analytic, individualistic attitude and exact comprehension. The image-oriented information processing of the Easterner implies holistic, synthetic, collectivistic attitude and fuzzy comprehension. As a corollary, we find that Eastern culture is by and large humanistic (or human-oriented) and fatalistic, including faith in fate or karma. At the same time, we see that Western culture is, in the same way, naturalistic (or nature-oriented) and prolific, depending on active position, creation and creativity.

Consequently, in her/his life, a typical Westerner is actively oriented at construction and transformation. At the same time, a typical Easterner is mostly absorbed in observation and imagination. As a corollary, we find that Eastern cultures are, as a rule, humanitarian (it means that they are aimed at acquiring knowledge about individuals and society); while Western cultures are, as a rule, technological (it means that they develop knowledge and practical tools for transformation knowledge for the benefit of people).

After seeing all these similarities and dissimilarities, it is natural to ask a question to what conclusions we come in such a way, or in other words, what is necessary to do aiming at a better society. The answer is not a simple one.

To provide some insight in the situation, I want to tell you the following story about one island. There was only one piece of land there that could produce some crops. Once upon the time, a man lived there. He cultivated this piece of land and harvested 1000 pounds of wheat. It was not very much but provided good living conditions for one person. Then another man came there. He got rid of the first man, seized the land, and began to harvest 1000 pounds of wheat himself. However, the first man returned and began to fight with the second man. They fought until one of them killed another. Nevertheless, the winner was so injured in the fighting that he died in a short time after his "victory".

Time passed, and another man came to this piece of land. He also began to cultivate it harvesting 1000 pounds of wheat. Then another man came and began fighting with the first one for this piece of land. However, this time, both men were cleverer than the first pair. They began to negotiate and agreed to divide the land into two equal parts. So, they lived there and each of them harvested 500 pounds of wheat. It was less only one than could have, but they were alive and had enough food.

Time passed, both men died, and for some time nobody lived at this place. Then another man came to this piece of land. He also began to cultivate it harvesting 1000 pounds of wheat. Then another man came but instead of fighting he suggested to work together on this land. The first man consented and they began to collaborate. The result was that together they harvested 4000 pounds of the crop because one knew a lot about plants and planting, while the second one had technical skills and could very well maintain and operate agricultural machinery. Thus, each of them had 2000 pounds of wheat and selling what they did not need for themselves, they were able to satisfy all their needs.

Now you can decide for yourselves what is better: opposition and fighting, separation and competition, or alliance and sincere collaboration.

Chapter 8

CONCLUSION

As many have found, we live in a bewildering world. We want to make sense of what is going on around us and this book suggests powerful tools for doing this. The suggested tools are symbolic structures and principles of finding and building such structures in all walks of our life. Only by structuring the world around us, we discern different things and construct a "world picture". We structure our activity and behavior to easier achieve our goals and improve our life.

Everything has a structure and finding these structures can help us not only to better understand our world but also to use the environment in a more efficient way. It means using the environment to our benefit but without destroying it, without acquiring little benefits today and bringing the world to a disaster tomorrow.

Thus, the main conclusion is that all people need to and must know about structures because structures reveal the essence of our life and the essence of our world. Even moral, life and love have structures.

In Chapter 4, structural approach to various phenomena called *structuralism* is described as an intellectual movement and scientific approach. It was very successful in the second half of the 20th century but in spite of all its achievements, structuralism was severely criticized and mostly rejected in humanities, philosophy and life sciences. However, this book shows that essential problems with this approach appeared due to insufficient knowledge of structures, too narrow understanding of the concept of *structure* and too simplistic models of structures and operations with them. Results obtained in this book eliminate such shortcomings suggesting more powerful and efficient *neostructuralism.* This direction may be also called *multistructuralism* because it asserts that complex objects have multiple structures. Taking into account all advantages described below, multistructuralism can accelerate the development of science, making it more people oriented.

In comparison with the conventional approach, the new conception of *structure* provides the following advantages.

1. In the conventional approach, it is assumed that any object has only one structure, the structure of this object, and this structure is comprehensive. In contrast to this, the results of this book demonstrate that an object, as a rule, has many representations as a system, while a system usually has many structures. Moreover, any object/system has structures of different types: inner structures, outer structures,

intermediate structures, etc. Thus, to describe an object in its entirety, it is necessary to find and use different structures.

2. A higher understanding and more advanced representation of structures are suggested based on mathematical models from (Burgin, 2010; 2011).
3. In contrast to the traditional approach, elements of structures are interpreted not only as elements but also as parts, components, modules and other constituents of systems, objects or entities. This eliminates unnecessary restrictions present in the traditional definition of a structure. In particular, this allows one to include (if necessary) the whole system (structure, object or entity) in its structure as an improper part of itself. This extended understanding allows researchers to consider all sets as some primitive structures.
4. While in a conventional structure, only relations between elements or parts are considered, the new concept of a structure also includes relations of parts to other entities and to themselves. In particular, this allows considering properties of parts as constituents of the structure. Namely, properties are intermediate relations. For instance, such a property as the *weight* of an individual is a relation between this individual and the scale in pounds, for example, by weighting, it is found that a woman weights 120 lb. When the scale in kilograms is taken, we say, for example, that this man weights 75 kg. Even more, properties of relations and relations between such properties also become constituents of the structure.
5. Inclusion of relations between relations is a new feature of the concept of a structure. The system of such relations forms the *structural hierarchy*, which is formalized in Appendix. In this hierarchy, elements and relations of the first level form the structure of the first level, elements and relations of the first and second levels form the structure of the second level and so on.
6. A new global model of the world as a whole called the Existential Triad is discussed and its relations with the structures of language, culture, sign and society are analyzed.
7. Given analysis of structural aspects of reality demonstrates that in contrast to the conventional approach where all phenomena and objects are separated into two groups – real and not real, actually there are many sorts and types of reality, which form an elaborate structure.
8. A general theory (metatheory) of structures, elements of which are presented in (Burgin, 2010; 2011), is further developed.

The results from this book allow us to suggest some directions for future research:

Direction 1. Further develop the general theory of structures, especially, its mereological aspects.

Direction 2. Study operations with structures in the context of the general theory of structures.

Note that a diversity of operations with specific structures, such as groups, topological spaces or manifolds, has been constructed and explored in mathematics, while algebras, i.e., mathematical systems with operations, have been elaborated for many of these structures. Various operations with data structures have been introduced and studied in information technology (databases, software engineering, etc.), bringing forth algebras of data structures, such as relational algebras (Codd, 1990). Operations with arbitrary structures represented by graphs were studied in (Aizerman, et al, 1987). Basic operations with triadic relations and corresponding algebras are investigated in (Robertson, 2005). A variety of operations with such a basic structure as a named set or fundamental triad are constructed and analyzed in (Burgin, 2010b; 2011).

Direction 3. Construct computational models that work with general structures and their representations in the context of the general theory of structures.

Various existing computational models work with specific structures. The most general data structure was used by Kolmogorov (1953) in his wide-ranging model of algorithms and computation. Kolmogorov algorithms work with finite structures that arere represented by networks and called *data configurations*. As it is discussed in Chapter 1, discrete graphs give an especially transparent representation of binary structures. That is why they are very popular structures in research and applications. Rabin (1969) studied automata on infinite trees, which are a special type of graphs. Besides, the multidimensional structured model of parallel computations studies computations with such a popular structure as *generalized arrays* (Burgin and Karasik, 1976; Burgin, 1984). It is useful to know that generalized arrays can naturally represent graphs, data configurations and many other useful structures.

To conclude, we would like to emphasize that the three worlds from the Existential Triad of the World are sides or basic aspects of the whole world. However, there is an interesting problem whether all three worlds can exist only in connection with one another or some (all) of them have independent being. Contemporary science tells us that, at least, two of these worlds, material and mental, cannot exist without structures. Whether the Structural World has independent being is the great problem, which goes back to Plato and Aristotle.

The world model provided by the Existential Triad allows us to analyze various processes that go in science. For instance, physicists are trying to create a theory of everything. However, this is impossible because physics studies only physical, i.e., material systems. So, in the best case, physicists will be able to create a theory of everything on the level of physics as the basic scientific reflection of the material world. However, there are higher than physical levels of reality – biological, psychological and social. Thus, it becomes, for example, questionable that a physical theory can explain all processes and phenomena in society. At the same time, we already know that everything has a structure and, as a rule, not a single one. Thus, the general theory of structures is a theory of everything as its studies everything. Now this theory is only at the initial stage of its development and its development requires work of many researchers.

APPENDIX

FORMAL CONCEPTS OF STRUCTURE

While it is possible only to name substance because any attempt to further that naming immediately involves attachment of some structure, we can independently develop the concept of structure to a certain extent of precision. We begin (Appendix, Part A) with list of basic definitions from set theory, which are necessary for correct understanding of materials about structures. Then (Appendix, Part B) we consider existing mathematical definitions and interpretations of the concept *structure*. After this (Appendix, Part C), we further develop the general theory of structures, which may be also called a metatheory of structures from (Burgin, 2010; 2011).

The difference between a theory of structures and a metatheory of structures is that any mathematical theory, e.g., theory of groups, functional analysis or probability theory, is a theory of some kind of structures. At the same time, a general theory of structures (metatheory of structures) studies structures defined as an abstract concept. For instance, Bourbaki (1957; 1960) developed elements of a certain metatheory of structures. It is possible to consider the theory of categories as a kind of metatheories of outer structures (Goldblatt, 1984; Ore, 1935; 1936; Corry, 1996). Here we develop a comprehensive metatheory of structures, which encompasses other existing approaches and at the same time, does not involve unnecessary complexity as the approach of Bourbaki.

A. Basic Set-Theoretical Definitions and Notation

$\varnothing$ is the *empty set*.

If X is a set, then $r \in X$ means that r belongs to X or r is a member of X.

If X and Y are sets, then $Y \subseteq X$ means that Y is a *subset* of X, i.e., Y is a set such that all elements of Y belong to X.

The *union* $Y \cup X$ of two sets Y and X is the set that consists of all elements from Y and from X.

The *intersection* $Y \cap X$ of two sets Y and X is the set that consists of all elements that belong both to Y and to X.

The *union* $\bigcup_{i\in I} X_i$ of sets X_i is the set that consists of all elements from all sets $X_{i,}$ $i\in I$.

The *intersection* $\bigcap_{i\in I} X_i$ of sets X_i is the set that consists of all elements that belong to each set X_i, $i\in I$.

The *difference* $Y \setminus X$ of two sets Y and X is the set that consists of all elements that belong to Y but does not belong to X.

If X is a set, then 2^X is the *power set* of X, which consists of all subsets of X. The *power set* of X is also denoted by $\mathbf{P}(X)$.

If X and Y are sets, then $X\times Y = \{(x, y); x \in X, y \in Y\}$ is the *direct* or *Cartesian product* of X and Y, in other words, $X\times Y$ is the set of all pairs (x, y), in which x belongs to X and y belongs to Y.

Y^X is the set of all mappings from X into Y.

$$X^n = \underbrace{X \times X \times \dots X \times X}_{n}$$

Elements of the set X^n called the *Cartesian power* of X have the form $(x_1, x_2, \dots, x_n)$ with all $x_i \in X$ and are called *n-tuples*, or simply, *tuples*.

The *direct* or *Cartesian product* $X_1 \times X_2 \times \dots \times X_n$ of sets X_i $(i = 1, 2, 3, \dots, n)$ is the set of all n-tuples $(x_1, x_2, \dots, x_n)$ such that $x_i \in X_i$ $(i = 1, 2, 3, \dots, n)$.

The *direct* or *Cartesian product* $\prod_{i\in I} X_i$ of sets X_i is the set of all sets $\{x_n ; i\in I\}$ such that $x_i \in X_i$ for all $i \in I$.

A fundamental structure of mathematics is *function*. However, functions are special kinds of binary relations between two sets.

A *relation* T between sets X_i $(i = 1, 2, 3, \dots, n)$ is a subset of the Cartesian product $X_1 \times X_2 \times \dots \times X_n$. The number n is called the *arity* of T. and denoted by $\alpha(T)$.

A *relation* T between sets X_i $(i \in I)$ is a subset of the Cartesian product $\prod_{i\in I} X_i$. If I is an infinite set, then the arity $\alpha(T)$ of T is infinite.

A *binary relation* T between sets X and Y is a subset of the Cartesian product $X\times Y$. In this case, the set X is called the *domain* of T ($X = \mathrm{Dom}(T)$) and Y is called the *codomain* of T ($Y = \mathrm{CD}(T)$). The *range* of the relation T is $\mathrm{Rg}(T) = \{ y ; \exists\, x \in X\, ((x, y) \in T) \}$. The *domain of definition* of the relation T is $\mathrm{DDom}(T) = \{ x; \exists\, y \in Y\, ((x, y) \in T) \}$. If $(x, y) \in T$, then one says that the elements x and y are in relation T, and one also writes $T(x, y)$.

Binary relations are also called *multivalued functions* (mappings or maps).

Taking two binary relations T: $X \times Y$ and R: $Y \times Z$, it is possible to build a new binary relation $R \circ T \subseteq X \times Z$ that is called the *sequential composition* of binary relations T and R and is defined by the rule $R \circ T = \{ (x, z); \exists\, y \in Y\, ((x, y) \in T \,\&\, (y, z) \in R) \}$.

A *preorder* (also called *quasiorder*) on a set X is a binary relation Q on X that satisfies the following axioms:

1. Q is reflexive, i.e. $x\mathrm{Q}x$ for all x from X.
2. Q is transitive, i.e., $x\mathrm{Q}y$ and $y\mathrm{Q}z$ imply $x\mathrm{Q}z$ for all $x, y, z \in X$.

A *partial order* is a preorder that satisfies the following additional axiom:

3. Q is antisymmetric, i.e., $x\mathrm{Q}y$ and $y\mathrm{Q}x$ imply $x = y$ for all $x, y \in X$.

A *strict* also called *sharp partial order Q* is a preorder that is not reflexive, is transitive and satisfies the following additional axiom:

4. Q is *asymmetric*, i.e., only one relation $x\mathrm{Q}y$ or $y\mathrm{Q}x$ is true for all $x, y \in X$.

A *linear* or *total order Q* is a strict partial order that satisfies the following additional axiom:

5. We have either $x\mathrm{Q}y$ or $y\mathrm{Q}x$ for all $x, y \in X$.

A set (class) X is *well-ordered* if there is a partial order on X such that any its non-empty subset has the least element. Such a partial order is called *well-ordering*.

An *equivalence* on a set (class) X is a binary relation Q on X that is reflexive, transitive and satisfies the following additional axiom:

6. Q is *symmetric*, i.e., $x\mathrm{Q}y$ implies $y\mathrm{Q}x$ for all x and y from X.

A *tolerance relation* is a binary relation that is reflexive and symmetric.

A *function* (also called a *mapping* or *map* or *total function* or *total mapping*) f from X to Y is a binary relation between sets X and Y in which there are no elements from X which are corresponded to more than one element from Y and to any element from X, some element from Y is corresponded. Often total functions are also called everywhere defined functions. Traditionally, the element $f(a)$ is called the image of the element a and denotes the value of f on the element a from X. At the same time, the function f is also denoted by $f: X \to Y$ or by $f(x)$. In the latter formula, x is a variable and not a concrete element from X.

A function (mapping) f from X to Y is an *injection* if the equality $f(x) = f(y)$ implies the equality $x = y$ for any elements x and y from X, i.e., different elements from X are mapped into different elements from Y. A function (mapping) f from X to Y is a *projection* also called *surjection* if for any y from Y there is x from X such that $f(x) = y$. A function (mapping) f from X to Y is a *bijection* if it is both a projection and injection.

A *partial function* (or *partial mapping*) f from X to Y is a binary relation between sets X and Y in which there are no elements from X which are corresponded to more than one element from Y. Thus, any function is also a partial function. Sometimes, when the domain of a partial function is not specified, we call it simply a function because any partial function is a total function on its domain.

A *multivalued function* (or *multivalued mapping*) f from X to Y is any binary relation between sets X and Y.

$f(x) \equiv a$ means that the function $f(x)$ is equal to a at all points where $f(x)$ is defined.

Two important concepts of mathematics are the domain and range of a function. However, there is some ambiguity for the first of them. Namely, there are two distinct meanings in current mathematical usage for this concept. In the majority of mathematical areas, including the calculus and analysis, the term "domain of f" is used for the set of all values x such that $f(x)$ is defined. However, some mathematicians (in particular, category theorists), consider the domain of a function $f: X \to Y$ to be X, irrespective of whether $f(x)$ is defined for all x in X. To eliminate this ambiguity, we suggest the following terminology consistent with the current practice in mathematics.

If f is a function from X into Y, then the set X is called the *domain* of f (it is denoted by Dom f) and Y is called the *codomain* of T (it is denoted by Codom f). The *range* Rg f of the function f is the set of all elements from Y assigned by f to, at least, one element from X, or formally, $\mathrm{Rg}\, f = \{ y ; \exists\, x \in X \; (f(x) = y) \}$. The *domain of definition* DDom f of the function f is the set of all elements from X that related by f to, at least, one element from Y is or formally, DDom $f = \{ x ; \exists y \in Y (f(x) = y) \}$. Thus, for a partial function $f(x)$, its domain of definition DDom f is the set of all elements for which $f(x)$ is defined.

Taking two mappings (functions) $f: X \to Y$ and $g: Y \to Z$, it is possible to build a new mapping (function) $gf: X \to Z$ that is called the *composition* or *superposition* of mappings (functions) f and g and defined by the rule $gf(x) = g(f(x))$ for all x from X. Note that the composition (superposition) of mappings/functions f and g coincides with the sequential composition of these mappings/functions as binary relations.

For any set S, $\chi_S(x)$ is its characteristic function, also called *set indicator function*, if $\chi_S(x)$ is equal to 1 when $x \in S$ and is equal to 0 when $x \notin S$, and $C_S(x)$ is its partial characteristic function if $C_S(x)$ is equal to 1 when $x \in S$ and is undefined when $x \notin S$.

If $f: X \to Y$ is a function and $Z \subseteq X$, then the restriction $f|_Z$ of f on Z is the function defined only for elements from Z and $f|_Z(z) = f(z)$ for each element z from Z.

If U is a correspondence of a set X to a set Y (a binary relation between X and Y), i.e., $U \subseteq X \times Y$, then $U(x) = \{y \in Y; (x, y) \in U \}$ and $U^{-1}(y) = \{x \in X; (x, y) \in U\}$.

An n-ary relation R in a set X is a subset of the nth power of X, i.e., $R \subseteq X^n$. If $(a_1, a_2, \ldots, a_n) \in R$, then one says that the elements $a_1, a_2, \ldots, a_n$ from X are in the relation R.

B. Conventional Theoretical Definitions of Structure

Looking at the ordinary usage of the word *structure*, we see that on the intuitive level, structure means a complex of objects and relations comparable with one another in principle. There are more exact definitions. For instance, Grossmann (1990) defines a structure in the following way.

Definition B1. A *structure* is (a representation of) a complex entity that consists of parts in relations to each other.

This definition implies that two structures are identical if and only if: (a) their *nonrelational parts* are the same; (b) their *relational parts* are the same; and (c) corresponding parts stand in corresponding relations. Note that here we once more encounter named sets. Namely, a structure is a named set (see Chapter 1) in which the support and set of names consist of system parts.

In addition, identity of structures, according to this definition, is a named set in which the support consists of non-relational parts, the set (component) of names consists of relations parts, and the naming correspondence establishes relations between all parts of one structure and all parts of the other structure. Thus, we have the following *structure identity triad*:

correspondence where identical parts
stand in identical relations
identity of non-relational parts ⟶ identity of non-relational parts

A similar definition of a mathematical structure is used in logic and mathematics (see, for example, (Robinson, 1963; Yaglom, 1980)).

Definition B2. A *mathematical structure* is an abstract (symbolic) system

$$A = < M; R_1, R_2, \ldots, R_n >$$

where M is a set and $R_1, R_2, \dots, R_n$ are relations in this set.

Note that this understanding of structure implies a direct representation of a structure by a named set in which the support and set of names are both equal to M, while the naming relation is the disjoint union of relations R_1 , R_2 , … , R_n . A more exact representation is obtained by means of labeled named sets when elements (ties) from different relations have different labels.

This definition of a mathematical structure is relatively simple and intuitive. However, there are some problems with this definition. First, the definition of a system in the general system theory is a particular case of Definition B2 (see, for example, (Mesarovic and Takahara, 1975)). This makes systems particular cases of structures, while as we know, material systems are not structures but only have structures. Second, some authors (see, for example, (Mal'cev, 1970)) use the name model or relational system for the concept defined in Definition B2.

Physicists and other scientists understand mathematical structures in a similar way. For instance, Tegmark (2008) gives the following definition.

Definition B3. A *mathematical structure* is a system that consists of:

1. The set of entities that belong to the union $S = S_1 \cup S_2 \cup \dots \cup S_n$ of sets $S_1, S_2, \dots, S_n$.
2. Functions (functional relations) on these sets and their Cartesian products

$$R_k: S_{i1} \times S_{i2} \times \dots \times S_{it} \to S_n$$

Given any structure (system) R, we denote by El_R the set of all elements from R and by Rel_R the set of all relations from R.

In their informal definition of a mathematical structure, Bourbaki (1948) add one more condition - axioms.

Definition B4. A *mathematical structure* is a set with relations in this set and axioms that these relations satisfy.

It is necessary to remark that Bourbaki (1957; 1960/1968) also elaborated a formal definition of a general mathematical structure as a very abstract concept. We give here a short description of the formal definition from (Bourbaki, 1960) omitting formal details and expressions.

Bourbaki start with a finite sequence of ordered pairs of whole numbers, calling this sequence an *echelon construction scheme*. Then taking such a scheme S and n terms $E_1, E_2, E_3, \dots, E_n$ in a formal theory T that is stronger that a theory of sets, such as ZF, they build an *echelon construction* E of the scheme S on the sets $E_1, E_2, E_3, \dots, E_n$, which are taken as the building blocks of an inductive construction. Each step of this construction consists either of taking the Cartesian product ($E \times F$) of two sets E and F obtained in former steps or of taking the power set 2^E of one of these sets, e.g., E. Such an echelon construction of the scheme S is a sequence of terms in the theory **T** built according to the scheme S. Then taking (in the theory **T**) one more collection of sets $E_1', E_2', E_3', \dots, E_n'$ and a formal representation of a group of mappings f_i: $E_i \to E_i'$, $i = 1, 2, 3, \dots, m$, Bourbaki determine *canonical extensions* with the scheme S of the mappings f_i to a mapping of an echelon construction E of the scheme S on the sets $E_1, E_2, E_3, \dots, E_n$.

After this, Bourbaki define a *typification* T of letters $x_1, x_2, x_3, \ldots, x_n$ in **T**. After this they define the concept of a *transportable relation* with respect to the typification T. After this, Bourbaki define a *species of structures* Σ in the theory **T** as a text that is a combination of letters $x_1, x_2, x_3, \ldots, x_n, s$, terms in **T**, a typification $T\{x_1, x_2, x_3, \ldots, x_n, s\}$ of the letters x_1, $x_2, x_3, \ldots, x_n, s$ in **T** called the *typical characterization* of the species of structures Σ and a relation $R\{x_1, x_2, x_3, \ldots, x_n, s\}$ that is transportable with respect to T called the *axiom* of the species of structures Σ.

To define a "species of structure" Σ, one takes:

(1) n sets $E_1, E_2, \ldots, E_n$, as "principal base sets."
(2) m sets $A_1, A_2, \ldots, A_m$, the "auxiliary base sets", and finally
(3) a specific echelon construction scheme $S(X_1, X_2, \ldots, X_n, A_1, A_2, \ldots, A_m)$.

All these auxiliary constructions and definitions allow Bourbaki to define a *structure of species* Σ, taking terms $E_1, E_2, E_3, \ldots, E_n$ in the theory **T** as principal base sets.

Definition B5. A term U in the theory **T** is called a *structure of species* Σ if the relation

$$T\{E_1, E_2, E_3, \ldots, E_n, U\} \,\&\, R\{E_1, E_2, E_3, \ldots, E_n, U\}$$

is a theorem in **T**.

The formal definition is essentially much more complex and much longer than their informal definition, covering several pages in the book (Bourbaki, 1960). Besides, this definition is too abstract and complicated even for the majority of mathematicians, who prefer to use an informal notion of a mathematical structure or the definition where a structure is formalized as a set with relations in this set. As Corry (1996) writes, Bourbaki's concept of *structure* was, from a mathematical point of view, a superfluous undertaking. Even Bourbaki themselves did not use this concept in their later books of the series *Eléments de Mathématique* after they had introduced it in *Theory of Sets* (1960).

Besides, we can see that this definition is not only too complicated but is also counter-intuitive because Bourbaki describe a structure as a logical term, i.e., a logical formula from a set-theoretical language. However, a structure may be described or characterized by a formula but a formula can be only a very specialized (logical) structure.

In addition, specific logical terms are called sets in the process of defining a structure (Bourbaki, 1968, p.262). However, logical terms are names of sets in formalized set theories but they are not sets themselves.

For instance, according to Cantor (1895), a set is "*a combination in some whole M of definite well distinctive things m from our observation or thought. These things are called "elements" of the set M.*"

Thus, formulas are used as specific mathematical names of mathematical objects, such as structures (Burgin, 2011). We all know that in real life, there is a difference between objects and their names. A similar difference exists between mathematical objects and their formal or informal names.

A category provides its specific approach of characterizing and describing a mathematical structure of a given kind, namely, in terms of preservation of the structure of mathematical objects by mappings between them because category may be viewed as and, as a rule, is a system that consists of objects bearing a certain kind of structure together with mappings between such objects preserving that structure (Awodey, 1996). For instance, topological spaces and continuous mappings between them form a category. The great power of category theory is in exploitation of mappings between objects with the same structure to reveal their properties (see, for example, (Goldblatt, 1979)).

Bucur and Deleanu (1968) give an explicit categorical definition of a *set-theoretical structure*.

Let us consider a category **BEns** of sets as objects and their bijections as morphisms.

Definition B6. A covariant functor

$$F\colon \mathbf{BEns} \to \mathbf{BEns}$$

is called the *species of structure*.

Let us consider an object M from the category **BEns**.

Definition B7. An element $r \in \mathrm{F}(M)$ is called a *structure* of the type F on the set M.

These definitions show how it is possible to define structures in abstract categories and not only in the category of sets. Topoi provide even more possibilities for such constructions (Goldblatt, 1984).

C. Advanced Concepts and Elements of a Metatheory of Structures

To build a metatheory of structures, we separate two sorts of structures: *abstract structures* and *embodied structures*. An embodied structure is a structure of some system, while an abstract structure is a structure by itself.

Mathematical definitions of *structure* show that there are two types of structures: *set-theoretical structures*, which consist of basic elements and relations, and *mereological structures*, which are based on such an area as mereology (see, for example (Leśniewski, 1916; 1992; Leonard and Goodman, 1940; Simons, 1987)) and consist of parts, which may be essentially connected, having common subparts. In this book, the emphasis is on set-theoretical structures because mathematical structures are, as a rule, set-theoretical.

A *set-theoretical abstract structure* consists of *inner elements*, which belong to this structure, *relations*, and perhaps *outer elements*, which do not belong to this structure but are associated with it.

Note that the concept of outer elements of a system depends on the context or environment in which this system is considered. For instance, the outer elements of such a system as an individual in his family are different from the outer elements of this individual in society.

Mathematics studies set-theoretical abstract structures. So, it is possible to find a diversity of examples of set-theoretical abstract structures in mathematics. For instance, a group, field, ring, topological space, filter, fiber, Euclidean space, metric space, and logical calculus can be treated as set-theoretical abstract structures.

Note that the given description of a set-theoretical abstract structure does not use the concept of a set or category. This shows that the concept of *structure* can be basic without references to other theoretical concepts, such as sets, systems or categories. This allows one to build a system of axioms for structures independently of other theoretical concepts, such as sets, systems or categories. To do this, it is necessary to consider structures of two types: structures with the emphasis on parts or elements and structures with the accent on relations. In such a way, it is possible to develop a formal theory of structures as it was done by Shapiro (1997). It is natural that axioms suggested by Shapiro are parallel to the axioms of set theory because sets are one of the basic types of mathematical structures.

The necessity in different concepts of a structure was emphasized by Landry (2007) who explained that any single formal framework for structure definition is unduly restrictive because different contexts demand different concepts of structure. For instance, considering only set-theoretical structures, we see that a set is not a structure because there are no relations between its elements (Carter, 2008). Thus, we also derive a rigorous definition for mereological structures because they give definite advantages for applications and in particular, allow one to explicate the structure of a set. To do this, we explain the mereological approach to the world representation, system theory and mathematics.

While the set theoretical approach portrays the world as a collection of sets (may be additionally structured) that consist of elements, *mereology* assumes that the world is a system of objects, which consist of parts. Whereas elements are necessarily separate from one another, parts may be essentially connected, having common subparts. For instance, fingers and palms are parts of hands and hands are parts of a body. However, we cannot say that fingers and palms are elements of hands and hands are elements of a body. The membership relation that is basic for set theory is dissimilar to the very notion of parthood that mereology is about. However, it is possible to treat sets as specific mereological objects.

Similar to a set-theoretical structure, a *mereological abstract structure* consists of *inner parts*, which belong to this structure, *relations*, and perhaps *outer parts*, which do not belong to this structure but are associated with it.

The roots of *mereology* can be traced back to the early days of philosophy, beginning with the Presocratic atomists, such as Democritus from Abdera (460-370 BCE) and Leucippus of Miletus (ca. 480 - ca. 420 BCE), and continuing throughout the texts of Plato, Aristotle, and Anicius Manlius Severinus Boëthius (ca. 480–524 or 525), who is commonly called Boethius. These philosophers conceived the part-whole relation as fundamental to ontology and metaphysics. It is also possible to find ideas related to mereology in the writings of medieval ontologists and scholastic philosophers, such as Garland the Computist (11th century), Jean Roscelin (ca.1045–ca.1120) also called Roscellinus, Pierre Abélard (1079–1142), Thomas Aquinas (ca. 1225–1274), Raymond Llull (1232–1315), Walter Burley (ca. 1275-1344), and Albert of Saxony (ca. 1320–1390), as well as in works of Joachim Jungius (1587–1657), Gottfried Wilhelm Leibniz (1646–1716) and Immanuel Kant (1724–1804).

A philosophical theory of parthood relations was developed by Franz Clemens Honoratus Hermann Brentano (1838–1917) and his pupil Edmund Gustav Albrecht Husserl (1859-1938) although in a format that makes it difficult to separate the analysis of mereological concepts from that of other ontologically relevant notions (such as the relation of ontological dependence). Brentano employed the concept of part variously as a tool of his philosophy, considering the fundamental conception of the relation part/whole for itself. This topic runs right through his basic philosophical thinking. at work in all his results The conception and application of part-whole relation as distinct from the set theoretical approach is also at work in all Husserl's results.

In a theoretical form, mereology was developed by Stanisław Leśniewski (1886-1939) as an ontological alternative to set theory (Leśniewski, 1916; 1992). Mereology was a part of his foundations of mathematics, which comprised three deductive theories: protothetic, ontology, and mereology. However, Leśniewski's works were largely inaccessible to the majority of the mathematical community because they were published in Polish. That is why the scientific community became acquainted with mereology in the form of the Calculus of Individuals developed by Leonard and Goodman (1940). From this time, ontologists and metaphysicians took a great interest in mereology. It was also used in computer science (see, for example, (Eschenbach, 1995)).

As Varzi writes (2007), mereology has the aim to recast set theory in nominalistic terms, using different concepts, such as *part*, *whole*, *part-whole relation*. The relation of being a "part of" is essentially different from relation of being an "element of" because the first relation is naturally transitive, while the second one is not. In a formal setting, the classical mereology assumes (Simons, 1987) that given an object *A*, its parts form a Boolen algbra with respect to the partial order induced by the relation "to be a part of", and with operations of sum and subtraction. Thus, the parts of an object are named by elements of a Boolen algebra, and we see that an object is a named set from the mereological perspective. In addition, researchers further developed mereology (see, for example, (Slupecki, 1958)).

An important direction in mereology is *mereotopology* (see, for example, (Smith, 1996; 1997)). In the same venue as mereology, mereotopology reconstructs set-theoretical topology, using extentionality, connection, fusion, principle of external contact, etc.

Ideas of mereotopology were first formulated by Whitehead (1919, 1920). He interpreted geometry of the physical spacetime as a system of "events" and of an "extension relation" between events. Later mathematicians formalized Whitehead's approach in a rigorous mathematical way, while logicians gave a logical representation of mereotopology, building first order theories with the domains consisting of "regions" (see (Casati and Varzi, 1999)).

Mereology and mereotopology are discussed in philosophy and used in such fields as formal ontology, knowledge representation, reasoning about knowledge, and spatial reasoning as an alternative to set theory. At the same time, logicians suggested axiomatizations of set theory based on a part-whole relation (Shoenflies, 1921; Foradori, 1937; Wegel, 1956).

In the mereological theory developed here, we split the traditional *parthood relation* into two new relations: the *part-whole relation*, which is denoted by E and meaning "is a part of the whole", and the *subpart relation* which is denoted by $\sqsubseteq$ and meaning "is a subpart of a part". Objects of this theory may acquire two roles: one role is *the whole* and another role is *a part*. To denote objects in the role *the whole*, we use bold characters, while to denote objects in the role *a part*, we use italic characters. Some objects may acquire both roles, while others may have only one role. Objects that may have only one role *the whole* are similar to classes in a set theory with two types of objects – sets and classes. While classes cannot be members, such wholes cannot be parts.

The construction of set-theoretical structures is based on the concept of a set. To formalize mereological structures, we introduce the concept of an arrangement. Here we give only qualitative definitions without a complete axiomatization.

Definition C1. A *simple arrangement A* often called an *object* is a collection of symbols called parts with the *part-whole relation* E also called the *meronomic relation* between symbol-parts and the symbol A, e.g., aEA means that a is a part of A.

Informally, arrangements are like sets, but while a set has elements, an arrangement has parts. Besides, whereas a set is structured by the *membership relation* $\in$, a simple arrangement is structured by the *part-whole relation* E and a general arrangement is structured by the *parthood relations* such as the part-whole relation, subpart relation and connection relation.

Here we do not demand reflexivity or transitivity of the part-whole relation, but preserve its antisymmetry. Besides, we add two more basic relations.

Definition C2. An *arrangement A* is a simple arrangement with two additional basic relations:

1. The *subpart relation* $\sqsubseteq$ between symbol-parts.
2. The *connection relation* $\bowtie$ between symbol-parts.

In many cases, these relations satisfy the following axioms:

1. Relations $c \sqsubseteq a$ and $c \sqsubseteq b$ for some c imply the relation $a \bowtie b$.
2. The relation $\bowtie$ is symmetric.
3. The relation $\sqsubseteq$ is transitive.

In contrast to sets where the inclusion relation is defined by means of the membership relation, all three relations in arrangements are independent in a general case. Note that in our approach, we split the traditional parthood relation into three more primitive relations: the *part-whole relation*, *subpart relation* and *connection relation*. The new approach is more flexible, allowing, for example, wholes that cannot be parts and parts that cannot be wholes. When the part-whole relation and subpart relation coincide, we obtain the traditional transitive parthood relation.

Mereology has been axiomatized in various ways based on the *parthood relation*. In many cases, it is assumed that the parthood relation partially orders its universe, meaning that everything is a part of itself (reflexivity), that a part of a part of a whole is itself a part of that whole (transitivity), and that two distinct entities cannot each be a part of the other (antisymmetry). However, in other approaches it is forbidden to anything ever being a part of itself. This makes the parthood relation irreflexive. Together with transitivity, it implies antisymmetry of the parthood relation.

A *binary relation* in an arrangement A is a set of pairs of parts of A. An *n-ary relation* in an arrangement A is a set of n-tuples of parts of A.

A *mereological abstract structure* consists of *inner parts* of this structure, *relations*, and perhaps *outer elements*, which do not belong to this structure but are associated with it.

A mereological abstract structure is built on an arrangement with a system of additional relations in the same way as a set-theoretical abstract structure is built on a set with a system of additional relations.

Definition B6 describes set-theoretical structures in the categorical language. In a similar way, it is possible to give a categorical definition of a *mereological structure*, taking a category of arrangements instead of the category of sets.

Embodied structures represent systems, i.e., they are structures of systems. Note that it is possible to treat any object, process or function as a system. However, to achieve a relevant generality, we start with the concept of an object. We define an *object* as anything that can be considered as something distinct from everything else. This is a very general concept, which is relative with respect to those who consider something as an object and to their point of view. It is possible that something, i.e., what is described or imagined, is treated as a regular object by some people, while others do not accept such a thing as an object. This is a common situation in the history of mankind in general and in the history of science in particular. For instance, when three great mathematicians of the 19th century, Carl Friedrich Gauss (1777-1855), Nikolai Ivanovich Lobachewsky (1793-1856), and Janosh Bolyai (1802-1860), discovered non-Euclidean geometries, for them, non-Euclidean geometries were mathematical objects, while all other mathematicians claimed that there is no such a mathematical object as a non-Euclidean geometry. When the great English physicist Paul Adrien Maurice Dirac (1902-1984) theoretically discovered a positive "electron", which was later called *positron*, all other physicists claimed that there is no such an object as a positive "electron".

To get a structure of an object A, we have, at first, to treat this object as a system R_A. Systems and their parts have three formalized representations. The most popular is the *set-theoretical representation* when a system R consists of elements, which form the set called the *element base* of R, and relations between these elements. The element base $B_e(R)$ of R allows obtaining all elements of R by application of permissible operations to elements from $B_e(R)$. When the element base contains all elements from the system, we come to the traditional approach to the system definition in system theory (Mesarovic and Takahara, 1975).

In a *mereological representation*, a system R consists of parts, which form the arrangement called the *mereological base* of the system R, and relations between these parts. The mereological base $B_m(R)$ of R allows obtaining all partts of R by application of permissible operations to parts from $B_m(R)$.

In the *categorical representation*, a system does not have elements or parts but only subsystems and quotient systems. Although it is possible to consider subsystems and quotient systems as parts of the system, the categorical definition of a structure is different (cf. Appendix B).

Here we consider an essentially more general definition of a system and its structure than the conventional definitions. In this context, a system is an object structured in a definite way. It is possible that different structures define the same system, while one system may have many structures.

Definition C3. An *embodied structure* is a structure of some system R. Its *inner elements* (*inner parts*) are elements (parts) of R. Its *outer elements* (*outer parts*) are elements (parts) of other systems related to R.

Thus, any structure of some system is an embodied structure.

It is possible to discern *intrinsic* and *ascribed* embodied structures. People get information about intrinsic embodied structures by observation, experiment and design for artificially created things. Often the observational and experimental information allows researchers to ascribe a symbolic structure to the object of observation or/and experiment. Thus, it is important to make a distinction between grounded and ungrounded ascribed structures. It is possible even to introduce a measure of groundedness for ascribed structures.

Analyzing system relations, we see that there are three basic sorts of relations:

- *Relations between inner elements/parts.*
- *Relations between outer elements/parts.*
- *Relations between inner elements/parts and outer elements/parts.*

Note that in the mereological approach, a system usually can be a part of itself.

These sorts of relations define three basic classes of embodied structures: *inner*, *outer* and *combined* structures, as well as four additional classes of embodied structures: *external*, *associated*, *extended* and *intermediate* structures.

Definition C4. a) An *inner*, or *internal*, structure $S_{IN}(R)$ of a system R contains only inner structural parts/elements, i.e., parts/elements of R, relations between these parts/elements, relations between these parts/elements and relations from $S_{IN}(R)$ and relations between relations from $S_{IN}(R)$.

b) All relations of an inner embodied structure $S_{IN}(R)$ of a system R are called *internal relations* of R.

Thus, the *inner structure* of a system represents this system as formed from elements (the set-theoretical approach) or parts (the mereological approach) and connections/relations, which are defined recursively. To show this, we introduce *levels* of relations and structures.

Structures hinge on relations. So, we separate three basic orders of relations:

- Relations between elements/parts have *the first order.*
- Relations between elements/parts and relations have *the second order.*
- Relations between relations have *the third order.*

It is necessary to remark that, as before, we consider two types of structures: set-theoretical structures and mereological structures. Mathematicians study only set-theoretical structures, usually assuming that any mathematical object is such a structure. However, in contrast to set-theoretical structures, when a set is not a structure (Carter, 2008), mereological structures allow one to explicate the structure of a set. Indeed, a mereological structure of a set has two kinds of parts – elements and subsets, as well as two relations: the membership relation for elements (i.e., parts of the first kind) and the inclusion relation for subsets (i.e., parts of the second kind).

As it was discovered, structures are characterized (defined) by relations of various levels, orders and sorts (Burgin, 2011). Here we give exact definitions for levels of relations.

Definition C5. a) An *internal relation* of the first level in a system M is a relation between the parts/elements from the system M.

b) There are two kinds of *internal relations* of the second level in a system M:

(1) relations between the parts/elements from M and internal relations of the first level in M;

(2) relations between internal relations of the first level in M.

c) There are two kinds of *internal relations* of the n-th level in a system M:

(1) relations between the parts/elements from *M*, internal relations of the (n - 1)-th level in *M* and perhaps internal relations of the (n - 2)-th and lower levels in *M*;

(2) relations between internal relations of the (n - 1)-th and internal relations of the (n - 1)-th and lower levels in *M*.

The level of relations stratifies the system of all relations and defines the *relational structural hierarchy* (see Figure C1). It is described in the following way. On the zero level, we have the set $\Im_0$ of all objects from this structure. On the first level of the structural hierarchy, we have the set $\Re_1$ of all relations between objects from this structure. On the second level, we have two sets: $\Re_{01}$ is the set of all relations between objects from this structure and relations from $\Re_1$, and $\Re_{11}$ is the set of all relations between relations from $\Re_1$. On the third level, we have five sets: $\Re_{01}$ is the set of all relations between object variables from this structure and relations from $\Re_1$, and $\Re_{11}$ is the set of all relations between relations from $\Re_1$. Then we can build the fourth level, the fifth level and so on. In such a way, we obtain the complete relational structural hierarchy CASH($\Im_0$) on the set of elements $\Im_0$.

Parts (substructures) of CASH($\Im_0$) are called relational structural hierarchies and denoted by ASH($\Im_0$). In general, a relational structural hierarchy ΛSH($\Im_0$) can grow as far as we want and even be infinite.

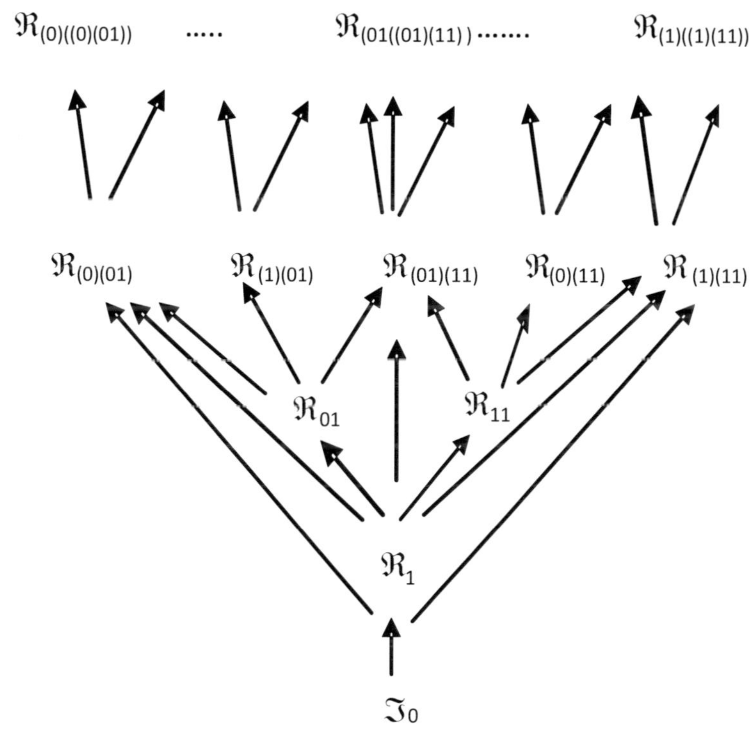

Figure C1. Five levels of the structural hierarchy.

We discern abstract relational structural hierarchies and embodied relational structural hierarchies, which are also called *relational structural hierarchies of systems.*

They are constructed in a similar way. Let us consider a *relational set-theoretical structural hierarchy* $SH_{ST}(M)$ for a system M with the element base Q. Note that M may be some structure and Q may include all elements from M.

On the zero level of $SH_{ST}(M)$, we have the set T_0 of all elements from Q. On the first level of the structural hierarchy $SH_{ST}(M)$, we have the set R_1 of all relations between object variables from this structure. On the second level of $SH_{ST}(M)$, we have two sets: R_{01} is the set of all relations between object variables from this structure and relations from R_1, and R_{11} is the set of all relations between relations from R_1. On the third level of $SH_{ST}(M)$, we have five sets: R_{01} is the set of all relations between object variables from this structure and relations from R_1, and R_{11} is the set of all relations between relations from R_1. Then we can build the fourth level, the fifth level and so on. In general, the relational structural hierarchy $SH_{ST}(M)$ of M can grow as far as we want and even be infinite.

Note that relations of the first order are exactly all relations of the first level. Relations of the second order can be on any level starting with the second level. The same is true for relations of the third order.

Exactly in the same way, it is possible to build the *mereological structural hierarchy* $SH_{ML}(M)$ for any system M with the mereological base R.

Example C1. As it is demonstrated in (Burgin, 2010), the clone of a universal algebra (Cohn, 1965) is an example of an algebraic structure that is presented as an infinite relational structural hierarchy.

Structural hierarchy $SH_{ST}(M)$ or $SH_{ML}(M)$ determines the *level of a structure* in this hierarchy.

Structure R *of the* (*strict*) *first level* of M consists of parts/elements from the base, relations between these parts or elements, i.e., relations of the first level. It means that the structure R belongs to the first level of the structural hierarchy $SH_{ST}(M)$ or $SH_{ML}(M)$. This is the conventional description of a structure in general, i.e., the conventional definition does not go above the first level of structures.

Structure R *of the second level* of M consists of parts/elements from the base and relations of the first and second levels, including, at least, one relation of the second level. It means that this structure R includes structures from the second level of the structural hierarchy $SH_{ST}(M)$ ($SH_{ML}(M)$) or their substructures and perhaps structures from the first level of the structural hierarchy $SH_{ST}(M)$ ($SH_{ML}(M)$).

Structure R *of the strict second level* of M consists of parts/elements from the base and relations of the second level. It means that this structure R belongs to the second level of the structural hierarchy $SH_{ST}(M)$ or $SH_{ML}(M)$ and does not belong to the first level of the structural hierarchy $SH_{ST}(M)$ or $SH_{ML}(M)$.

For an arbitrary $n > 1$, we have the following definition.

Structure R of the *level n* of M consists of parts/elements from the base and relations of the level n and smaller levels, including, at least, one relation of the level n. It means that this structure R includes structures from the n–th level of the structural hierarchy $SH_{ST}(M)$ ($SH_{ML}(M)$) or their substructures and perhaps structures from the lower levels of the structural hierarchy $SH_{ST}(M)$ ($SH_{ML}(M)$).

Structure R of the *strict level n* of M consists of parts/elements from the base and relations of the level n. It means that this structure R belongs to the n–th level of the structural hierarchy $SH_{ST}(M)$ ($SH_{ML}(M)$) and does not belong to the $(n - 1)$-th level of the structural hierarchy $SH_{ST}(M)$ ($SH_{ML}(M)$).

In such a way, we obtain a *relational structural hierarchy* of the system M, which may be infinite in a general case because n can go to infinity, and it is possible to define a structure as a structural hierarchy or its part.

In addition to the relational structural hierarchy, there is also a *compositional structural hierarchy* of structures. In it, some structures are parts or substructures of other structures. For instance, the structure of an organism is composed of the *structure of bones*, the *nervous structure*, the *muscular structure*, the *articulatory structure* and the *visceral structure*. Thus, the structure of an organism is higher in the compositional structural hierarchy than the structure of bones or the nervous structure, while the nervous structure is higher in the compositional hierarchy than the structure of the central nervous system, which in turn, is higher than the structure of a neuron.

The prominent philosopher of the 20th century Rudolf Carnap defines a *constructional system* of objects as a system in which the objects of each level are constructed from objects of more elementary levels of construction. The most elementary level of a constructional system is the level of basic objects, which include basic elements and basic relations. Carnap describes the basic elements of a constructional system as "elementary experiences," which are not constructed but which are immediately given as formal objects (Carnap, 1928). It is possible to see that the structure of a constructional system is compositional hierarchy of structures.

Example C2. Let us consider inner structures of an organism. They are organized in the following compositional hierarchy:

organism
↑
organ system
↑
organ
↑
tissue
↑
cell

Thus, the inner structure of an organism on the first (top) compositional level consists of organ systems with different relations. There are such organ systems as the *nervous system*, *circulatory system*, *digestive system*, *muscular system*, *respiration system*, and *skeletal system*.

The inner structure of an organism on the second compositional level consists of organs with different relations. The inner structure of an organism on the third compositional level consists of tissues with different relations. The inner structure of an organism on the fourth compositional level consists of cells with different relations.

An important example of a compositional structural hierarchy is described in (Cottrell and Pettifor, 2000). This hierarchy is studied in the field of material sciences, which is the branch of the science of complex structures. The authors consider four levels of this hierarchy: the *level of electronic structures*, the *level of atomistic* or *atomic structures*, the *microstructural level* and the *continuum level*. Physics studies electronic structures.

Chemistry studies atomistic structures. The microstructural level is in the scope of material sciences and the continuum level is encompassed by engineering. An important peculiarity of these levels is that although the structures of the higher level are composed from the structures of the lower level, their mathematical models use different structures and techniques that are not compositionally connected.

Remark C1. There are other ways to build structural hierarchies, which are considered elsewhere.

Let us formalize the concept of an outer relation introduced in Chapter 1.

Definition C6. a) An *outer embodied structure* $S_{OT}(R)$ of a system R is an inner structure of a system T in which R is only one of its inner parts/elements of the structure $S_{IN}(R)$ of the system T.

b) All relations of an outer embodied structure are called *outer relations*.

Thus, the *outer structure* of a system R is the inner structure of a supersystem that contains R as a part or element and an outer relation of R is the inner relation of a supersystem that contains R as a part or element.

For instance, taking a Turing machine T that is a node in a grid automaton G, we can contemplate the structure of this grid automaton as the outer structure of T (Burgin, 2005). In this case, ports are usually ignored and we consider grid automata without ports.

Note that the concept of an outer structure of a system depends on the context or environment in which this system is considered. For instance, the outer structure of an individual in his family is different from the outer structure of this individual in society.

It is possible to specify the inner structure of an object (system) by its outer structure, describing relations between objects with the same structure. This approach was used by Ore (1935; 1936) for formalizing the concept of an algebraic structure and further developed in category theory aimed at formalizing the concept of a mathematical structure (see (Corry, 1996)). Both approaches – inner and outer - are based on the idea that the structural nature of algebras in the case of Ore and mathematical structure in general in the case of categories can best be understood by entirely ignoring the existence of the elements of a given mathematical, e.g., algebraic, domain and concentrating on the relationships between the domain and its subdomains.

Example C3. A mathematical category is an outer structure of any of its objects. In this structure, elements are connected by morphisms. For instance, the category of groups is an outer structure of a group. In this structure, elements are connected by homomorphisms.

Example C4. An *abstract data type* is a data structure that is defined indirectly by the operations that may be performed on it, and the mathematical properties of those operations (including their space and time cost). Thus, an abstract data type is defined by its inner and intermediate structures.

Example C5. One more example is given by the concept of an object in object-oriented programming. We remind here that *object-oriented programming* (OOP) is a programming technique (paradigm) using "objects" to design applications and computer programs (Abadi and Cardelli, 1996).

Object-oriented programming is a type of programming in which programmers define not only the data type of a data structure, but also the types of operations (functions) that can be applied to the data structure. In this way, the data structure becomes an *object* that includes both data and functions. Thus, an *object* becomes a software structure that consists of a related *state* represented by data fields and *behavior*, which includes *methods* or *actions* or *processes* together with their interactions. Software objects are often used to model the real-world objects from everyday life, science, engineering, medicine, etc.

They all have *state* and *behavior*. Dogs have state with the fields *name*, *color*, *breed*, *weight*, *age*, *hungry*, *satisfied*, etc. and behavior with actions/processes *barking*, *fetching*, *wagging tail*, etc. Bicycles also have state (current gear, current pedal cadence, current speed) and behavior (changing gear, changing pedal cadence, applying brakes).

A software object stores its state in *fields* (variables in some programming languages) and exposes its behavior through *methods* (functions in some programming languages).

In addition, programmers can create relationships between one object and another. For example, objects can *inherit* characteristics from other objects.

One of the principal advantages of object-oriented programming techniques over procedural programming techniques is that they enable programmers to create modules that do not need to be changed when a new type of object is added. A programmer can simply create a new object that inherits many of its features from existing objects. This makes object-oriented programs easier to modify.

To perform object-oriented programming, one needs an *object-oriented programming language* (OOPL). Java, C++ and Smalltalk are three of the most popular object-oriented programming languages, and there are also object-oriented versions of Pascal. There is an extended literature where it is possible to read more about object-oriented programming and its applications.

It is possible to build the structural hierarchies of outer structures in the same way as it is done above for internal structures. There are two kinds of such hierarchies: the relational structural hierarchy and compositional structural hierarchy.

Example C6. An infinite compositional structural hierarchy emerges in the realm of categories. As Manin (2004) writes, "Categories themselves form objects of a larger category *Cat* morphisms in which are functors, or "natural constructions" like a (co)homology theory of topological spaces. However, functors also form objects of a category. Axiomatizing this situation, we get a notion of *2-category* whose prototype is *Cat*. Treating 2-categories in the same way, we get 3-categories etc."

Example C7. Let us consider outer structures of an organism. They are organized in the following compositional system hierarchy:

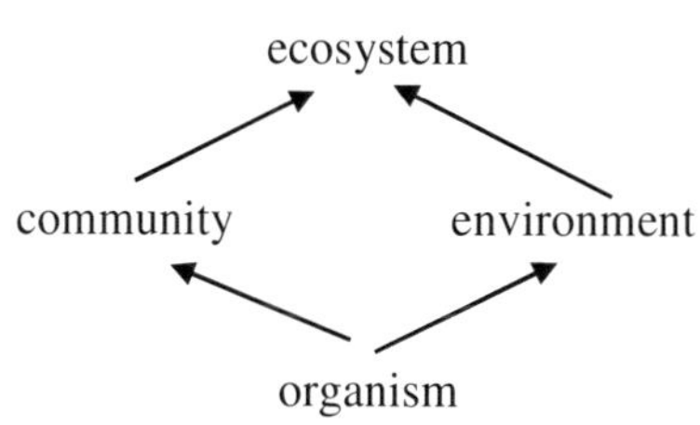

Thus, one outer structure of an organism, or better to say, an individual with this organism, on the first compositional level is the structure of the community to which this individual (organism) belongs. Another outer structure of an individual (organism) on the first compositional level is the structure of the environment of this individual (organism). An outer structure of an individual (organism) on the second compositional level is the structure of the ecosystem to which this individual (organism) belongs.

There are important relations between inner and outer structures.

Definition C7. An outer structure $S_{OT}(M)$ of a system M is *correlated* with an inner structure P (type T of inner structures) if all elements/parts in the structure Q have the inner structure P (inner structures of all elements/parts from $S_{OT}(M)$ belong to the type T).

For instance, an (algebraic) category is constructed in such a way that its structure an outer structure of any of its elements (objects) and is correlated with the inner structure of these objects.

While an outer structure of a system R characterizes outer relations of the whole system R, an external structure of a system R characterizes outer relations of the of the parts/elements from the system R.

Definition C8. a) An *external structure* $S_{ET}(R)$ of a system R represents this system as formed from elements/parts of this system and some relations of these elements/parts to elements/parts from other systems, as well as to relations from other systems, and some relations between all these relations.

b) All relations of an external embodied structure $S_{ET}(R)$ of R are called *external relations* of R.

Example C8. Taking a module M over a ring R, which acts on M, we see that inner relations in M, i.e., relations between elements from M, are defined by operations of addition and subtraction. However, there are also relations between elements from M and elements from R, which are defined by operations of multiplication of elements from M by elements from R. These relations are external.

Example C9. Taking a polygon P over a semigroup T, which acts on P, we see that there are relations between elements from P and elements from T, which are defined by operations of multiplication of elements from P by elements from T. These relations are external.

It is possible to build relational structural hierarchies of external structures in the same way as it is done above for internal structures.

Definition C9. a) An *external relation* of the first level in a system M is a relation between the elements/parts of the system M and elements/parts from other systems. These systems are called *scales* of the corresponding external structure $S_{ET}(M)$ of M.

b) There are four kinds of *external relations* of the second level in a system M:
   (1) relations between the parts/elements from M and external relations of the first level in M;
   (2) relations between external relations of the first level in M;
   (3) relations between the parts/elements from M and internal relations of the first level in the scales of the external structure $S_{ET}(M)$ of M;
   (4) relations between external relations of the first level in R and internal relations of the first level in the scales of the external structure $S_{ET}(M)$ of M;

c) There are four kinds of *external relations* of the n-th level of a system M:
   (1) relations between the parts/elements from M, external relations of the $(n - 1)$-th level in A and perhaps external relations of the $(n - 2)$-th and lower levels in M;
   (2) relations between external relations of the $(n - 1)$-th and external relations of the $(n - 1)$-th and lower levels in M.
   (3) relations of the parts/elements from M with internal relations of the $(n - 1)$-th level in the scales of the external structure $S_{ET}(M)$ of M and perhaps external relations of the $(n - 2)$-th and lower levels in M, as well as in the scales of the external structure $S_{ET}(M)$ of M;

(4) relations between external relations of the $(n - 1)$-th level in M and internal relations of the $(n - 1)$-th and lower levels in the scales of the external structure $S_{ET}(M)$ of M, as well as in the scales of the external structure $S_{ET}(M)$ of M.

Example C10. Abstract properties in the sense of (Burgin, 1985; 1990a; 2011) are examples of external relations of the first level. Relations between abstract properties are examples of external relations of the second level.

It is possible that defined above scales of a structure Q of a system M also have structures. Adding these structures to the initial structure of M, we obtain an extended structure of the system M.

Definition C10. An *extended structure* $S_{EX}(M)$ of a system M consists of an inner structure $S_{IN}(M)$ of M and of inner structures of the scales of the external structure of M.

Example C11. Taking a module M over a ring R, we have the extended structure of a module, which consists of the internal structure of the module M and the structure of the ring R.

It is possible to build the relational structural hierarchy of extended structures in the same way as it is done above for internal structures.

Taking the outer part of an extended structure of a system, we obtain an associated structure of this system.

Definition C11. a) An *associated structure* $S_{AT}(M)$ of a system M consists of inner structures of the scales of the external structure of M.

b) All relations of an associated embodied structure $S_{AT}(M)$ of M are called *associated relations* of M.

Example C12. If M is a left module over a ring R, then the associated structure of M is the inner structure of the ring R.

Note that an object and even a system may have different extended and associated structures.

Example C13. Let us consider a left module over a ring R. Then the standard extended structure $S_{EX1}(M)$ of M consists of the inner structure $S_{IN1}(M)$ of M as an abelian group and the inner structure of the ring R, while the associated structure of M is the inner structure $S_{IN}(R)$ of the ring R.

At the same time, it is possible to any element r from R, to assign the relation $T_r \subseteq M \times M$ that represents the unary operation (action of r) m_r: $M \to M$ where $m_r(a) = ra$ is the multiplication by r. In this case, the inner structure $S_{EX2}(M)$ of M consists of the structure of an abelian group and of all relations T_r for $r \in R$. This is also the extended structure $S_{EX2}(M)$ of M, while in this representation, M does not have the external structure and the associated structure.

Note that while being different, these structures define the same mathematical object M.

Definition C12. a) An *intermediate structure* $S_{IM}(M)$ of a system M is an inner structure $S_{IN}(T)$ of a system T in which M is one of inner parts/elements of the structure $S_{IN}(T)$ and all other parts/elements of T are connected to M, i.e., the system M is considered as a whole with its (specified) relations with and connections to other systems.

b) All relations of an intermediate embodied structure $S_{IM}(M)$ of M are called *intermediate relations* of M.

We can see that an intermediate structure $S_{IM}(M)$ of a system *R* is the part of some outer structure $S_{OT}(M)$ of *M* that contains only relations that include *M* and systems that are related to *M*. Thus, all intermediate relations are also outer relations.

Example C14. Let us once more consider a Turing machine *T* that is a node in a grid automaton *G* (see, (Burgin, 2005)). We generate a subautomaton *H* of *G* by taking all edges that directly connect *T* to other nodes and these nodes. The structure of the grid automaton *H* is an intermediate structure of *T*. In this case, ports are essential and we consider grid automata with ports.

Example C15. In (Hsiao, *et al*, 2008; 2008a), intermediate structures of objects are represented by object schemas, which are special cases of static schemas in the sense of mathematical schema theory developed by Burgin (2006; 2010a).

An *object schema* is a network of actions that could be taken on an object, tied to effects that would result from each action.

Another representation of an object schema is in a form of the set (sum) of the object *affordances*, i.e., actions that can be taken with respect to the object and what they provide for an agent.

Object schemas are used by Hsiao, *et al* (2008; 2008a) for physically grounded natural language interpretation by robots that react appropriately to unanticipated physical changes in the environment and dynamically acquire new information pertinent to ongoing tasks. A robot encodes its beliefs about physical objects in its environment using object schemas and collections of coupled processes responsible for sensorimotor interaction.

All considered types of structures are synthesized in the concept of a combined or mixed embodied structure.

Definition C13. A *combined* or *mixed embodied structure* $S_{MX}(M)$ is a structure of a system *M* that can have parts/elements from *M* and from other systems, as well as relations between these parts/elements, relations between these parts/elements and relations from *M* and relations between relations from *M*.

The concept of a combined structure integrates inner and external structures and allows us to consider more general structural hierarchies. In particular, adding scales of properties *P* of elements, components and parts of *M* to system components of the structure, it is possible to include abstract properties that represent properties *P* in the relational structural hierarchy $SH_{ST}(M)$ or $SH_{ML}(M)$. Thus, we may assume, that the relational structural hierarchy $SH_{ST}(M)$ ($SH_{ML}(M)$) already contains all abstract properties that represent considered properties *P* of elements (components and parts) of *M*.

Given a structure *R*, it is possible to consider substructures of the structure *R* (see Definition C27), as well as properties and relations between substructures. In this context, any substructure *W* of a structure *R* has a representation by relations in *R*. Thus, it is possible to represent relations between substructures by relations between relations in the structural hierarchy $SH_{ST}(R)$ or $SH_{ML}(R)$.

Definition C14. A structure *Q* of a system *M* is a structure of *M* of the level *n* if *Q* does not contain relations that have the level higher than *n*.

When a system *M* is a part of itself, intermediate relations of *M* are special external relations of *M* and thus any intermediate structure of *M* is a substructure an external structure of *M*. As combined structures include inner and external system relations, we have the following result.

Proposition C1. If a system *M* can be a part of itself, then:

a) All inner, intermediate and external relations of M are special cases of combined relations of M.

b) Any inner, intermediate and external structure of M is a substructure of the combined structure of M.

One and the same structure can have different representations. To study the system of all possible symbolic representations of structures, we introduce types of symbolic structure representations:

1. *Algorithmic/procedural* representations.
2. *Nominalistic/descriptive* representations.
3. *Combinatorial/conceptual* representations.

Generative schemas introduced in (Burgin, 2010) for structure definition and formal grammars (see, for example, (Burgin, 2005)) are examples of algorithmic/procedural representations of structures.

Logical formulas used in (Bourbaki, 1960) for structure definition are examples of nominalistic/descriptive representations of structures. Another example of a nominalistic/descriptive representation is portrayal of a structure of the first level in the form of a list of its elements/parts and relations in the form of sets of tuples (see Appendix A) of the structure's elements/parts.

A combinatorial/conceptual representation depicts structures as classes of structurally equivalent objects. Language partitions studied in (Burgin and Burgina, 1982; 1982a) are examples of combinatorial/conceptual representations of structures.

There are also three modes of algorithmic/procedural representations:

1. *Generation* from initial information.
2. *Explication* from a system description.
3. *Recognition/selection*.

Besides, there are three layers of algorithmic/procedural representations:

1.1. An *analytic generation* of a structure by a function. For instance, a function can be defined by a formula or a static schema in the sense of (Burgin, 2010) and generation goes by substitution in this formula or a static schema.

1.2. An *algorithmic generation* of a structure by an algorithm. For instance, constructive rules are given for structure generation.

1.3. A *generation by an automaton/machine*. For instance, a structure is generated by a Turing machine.

In addition, there are three layers of nominalistic/descriptive representations:

1.1. A formal *symbolic description* of a structure, e.g., by a formula.

1.2. An *iconic* or *visual description* of a structure, e.g., by a graph or flowchart.

1.3. An informal *description by a natural language*. For instance, a triangle is a geometrical object that has three points called *vertices* connected by segments called *edges* or *sides*.

On the social level, structures are represented by systems of signs and symbols and are themselves symbols of things, systems and processes (see Chapter 6).

Each system has *inherent structural* representations. At the same time, inherent structural representations are (more or less) revealed by *reflected structural* representations instantiated in other systems.

There are three main types of structural representations:

1. *Material* in the Physical World.
2. *Symbolic* in the Mental World.
3. *Intrinsic* in the Structural World.

Types 2 and 3 of structural representations are abstract systems. In the physical world, abstract systems are represented by such elements as elementary linguistic objects, e.g., letters and digits, and relations that are subsets of Cartesian powers of sets of these elements.

It is possible to discern three kinds of material structural representations:

- *Representation by things*, e.g., by diagrams on the paper or on computer screen or by text printed on the paper or on computer screen.
- *Representation by processes.*
- *Representation by actions* or *events.*

Some researchers suggested a conjecture that all structural representations of objects mirror inherent structures of these objects. For instance, the symbolic inner structure of a sentence in English is a linear sequence of symbols each of which denotes a word.

For instance, Russell (1927) suggested the adoption of the following principle.

Mirroring Relations Principle. Relations between percepts correspond to relations between their non-perceptual causes in a manner which preserves their logico-mathematical properties.

According to Russell, this principle implies that from the structure of our perceptions, it is possible "infer a great deal as to the structure of the physical world, but not as to its intrinsic character" (Russell, 1927). In other words, he argues that all researchers can assert is that the structure of our perceptions is in some sense isomorphic to the structure of the physical world.

It is possible to separate three kinds of symbolic structural representations:

- *Formal representation*, e.g., by mathematical formulas.
- *Semiformal representation*, e.g., by texts in natural languages combined with mathematical formulas.
- *Informal representation*, e.g., by texts in natural languages.

Besides, it is necessary to discern pattern structures and individual structures.

Definition C15. A *pattern structure* is a structure common to several systems.

For instance, structure of a triangle is the same for a multitude of triangles.

It is possible to show that a structure in the sense of Bourbaki is a special kind of pattern structures defined in this book because axioms are kinds of relations of higher levels defined by formulas. However, not all structures in the sense of this book can be represented by structures in the sense of Bourbaki. For instance, it is impossible to represent structures where a system is considered as part of itself. Indeed, according to the axioms of set theory a set cannot be an element of itself and Bourbaki's construction is based on sets (Bourbaki, 1960).

Pattern set-theoretical structures have three basic representations:

1. A class of structurally equivalent objects, i.e., in the form of a combinatorial/conceptual representation.
2. A schema, i.e., in the form of an algorithmic/procedural representation.
3. A set of parts/elements (the base) and generative rules, i.e., in the form of a nominalistic/descriptive representation.

Definition C16. An *individual structure* is a specific structure of an individual object.

For instance, taking a definite triangle and determining its structure so that it would define a unique triangle, we obtain an individual structure of this triangle. However, when it is possible to make indistinguishable copies of an object, then the individual structure (of this object) becomes the pattern structure (of all copies). For instance, it is possible to make copies of formal (mathematical) triangles and their physical images, as well as of different words and texts. Thus, their individual structures become pattern structures.

In their definition, Bourbaki deal only with pattern structures, ignoring individual structures, because they assume that mathematics studies patterns. However, in principle, mathematics can also study individual structures.

These concepts show that there are more general and less general structures. Namely, a structure R is *more abstract* than a structure Q if R has the same elements as Q but only a part of relations from Q. Note that, in particular, in a more abstract structure, elements have a smaller amount of properties than in a *less abstract* structure because properties are external relations. For instance, the structure of a particular triangle ΔABC is less abstract than the general structure of a triangle.

Bourbaki suggest that there are three basic types of mathematical structures: *order structures*, *algebraic structures*, and *topological structures* (Bourbaki, 1960). The renowned psychologist and philosopher Jean Piaget (1896-1980) called them "mother structures", arguing that all mathematical formations are transformations of these "mother" structures (Piaget, 1964). Here we extend this classification of Bourbaki because these three classes do not encompass all mathematical structures. For instance, logical structures or algorithmic structures do not belong to any of these classes.

At first, let consider the classification of Bourbaki.

Order structures are described in Section A1. Here we give definitions of algebraic structures (see, for example, ((Kurosh, 1963; Cohn, 1965; Malcev, 1970)) and topological structures (see, for example, (Kuratowski, 1966; 1968)). The most general concept of the first type is an algebraic system.

Definition C17. An *algebraic system* is a structure $A = (X, \Omega, R)$ that consists of: a non-empty set X called the *carrier* or the *underlying set* of A and elements of which are called the elements of A; a family Ω of algebraic operations, which are mappings $\omega_i: X^{n_i} \to X$ $(i \in I)$; and a family R of relations $r_j \subseteq X^{m_j}$ $(j \in J)$ defined on X. The pair (Ω, R) is called the type of the algebraic system A.

Note that the concept of an *algebraic system* is based on the structure *named set* and it is possible to represent any algebraic system as a named set (Burgin, 2011). We also remind some basic concepts from the theory of algebraic systems (Malcev, 1970) because they are useful for the general theory of structures.

Definition C18. The non-negative integers n_i and m_j are called the *arities* of the respective operations ω_i and relations r_j, i.e., $\alpha(\omega_i) = n_i$ and $\alpha(r_j) = m_j$.

Definition C19. When ω is an operation from Ω with arity n, then the image $\omega(a_1, a_2, \ldots, a_n)$ of the element $(a_1, a_2, \ldots, a_n)$ from X^n under the mapping $\omega: X^n \to X$ is called the *value* of the operation ω at for elements $a_1, a_2, \ldots, a_n$. Usually it is denoted by $a_1a_2 \ldots a_n\omega$.

In a general case, some operations from Ω may be partial mappings. For instance, division in such a universal algebra (algebraic system) as a field is not defined for 0, i.e., it is impossible to divide by 0.

Definition C20. The operations from Ω and the relations from R are called *basic* or *primitive*.

Using basic operations, it is possible to build many different derivative operations in the algebra A. The pair of families $(\{n_i; i \in I\}; \{m_j; j \in J\})$ is called the *type* of the algebraic system A. Two algebraic systems A and A' have the *same type* if $I = I'$, $J = J'$, and $n_i = n_i'$ and $m_j = m_j'$ for all $i \in I$, $j \in J$. Here we see one more named set in which types are corresponded to algebraic systems.

Definition C21. An algebraic system $A = (X, \Omega, R)$ is called a *universal algebra* or simply, an *algebra* if the set R of its basic relations is empty. The set Ω is called the type of the universal algebra A.

Classical algebraic systems are groups, rings, linear (vector) spaces, linear algebras, lattices, ordered sets, ordered groups, etc. Groups, rings, modules, fields, linear spaces, linear algebras, lattices, and semigroups are classical universal algebras.

Definition C22. An algebraic system $A = (X, \Omega, R)$ is called a *model* (which is one of the basic concepts of logic) or a *relational system* if the set Ω of basic operations is empty.

Topological structures are defined in a different way.

Definition C23. A *topology* in a set X is a system $O(X)$ of subsets of X that are called open subsets and satisfy the following axioms:

T1. $X \in O(X)$ and $\varnothing \in O(X)$.

T2. For all A, B, if $A, B \in O(X)$, then $A \cap B \in O(X)$.

T3. For all A_i, $i \in I$, if all $A_i \in O(X)$, then $\cup_{i \in I} A_i \in O(X)$.

A set X with a topology in it is called a *topological space*.

Topology in a set can be also defined by a system of neighborhoods of points from this set. In this case, a set is *open* in this topology if it contains a standard neighborhood of each of its points. For instance, if a is a real number and $t \in R^{++}$, then an open interval $O_t a = \{ x \in R; a - t < x < a + t \}$ is a standard neighborhood of a.

If X is a subset of a topological space, then $Cl(X)$ denotes the closure of the set X.

In many interesting cases, topology is defined by a *metric*. Mappings of topological (metric) spaces that preserve the topological (metric) structure are called *continuous*.

However, these three "mother" mathematical structures of Bourbaki do not encompass all mathematical structures and we suggest a new classification. According to the type of basic relations, i.e., relations between elements, it is possible to separate three basic categories of structures:

1. If n is a natural number, then *n–ary structures* have only relations of arity n and may be less than n.
2. *Finitary structures* have only relations of finite arity but the arity is not bounded.
3. *Infinitary structures* have relations of infinite arity.

Three "mother" mathematical structures of Bourbaki are particular cases of these types. Namely, order structures are binary structures. Algebraic structures are finitary structures. Topological structures are infinitary structures.

In what follows, we consider only set-theoretical structures and their nominalistic or descriptive representations by means of formal symbolic expressions. Two other types, algorithmic/procedural representations and combinatorial/conceptual representations, are considered and analyzed in (Burgin, 2010; 2011).

We remind that a set-theoretical structure Q is a set with relations of different types and levels. For simplicity, we consider here only inner relations. General structures with other types of relations are considered elsewhere. In a set-theoretical structure Q, $\mathbf{El}_Q$ is the collection of elements from Q, and $\mathbf{Rel}^k_Q$ is the collection of relations of the level k from Q.

All sets $\mathbf{Rel}^k_Q$ are *graded*, i.e., each set $\mathbf{Rel}^k_Q$ is the disjoint union $\biguplus_j \mathbf{Rel}^{k,j}_Q$ where $\mathbf{Rel}^{k,j}_Q$ is the collection of all relations from $\mathbf{Rel}^k_Q$ that have arity j.

Basic concepts of the metatheory of structures are: a structural homomorphism, substructure, and structural isomorphism (equivalence), as well as structural homomorphisms of different levels and substructures of different levels. Let us consider some basic results of the metatheory of structures.

Taking two structures Q and R, we define structural homomorphisms.

Definition C24. A *structural homomorphism* $F: Q \to R$ of the level k is a sequence $F = (f_0, f_1, \dots, f_k)$ where:

f_0: $\mathbf{El}_Q \to \mathbf{El}_R$ is a mapping;

f_1: $\mathbf{Rel}^1_Q \to \mathbf{Rel}^1_R$ is a *graded mapping*, i.e., a mapping that preserves the arity of relations, such that it satisfies the following conditions:

(1) If $r \in \mathbf{Rel}^1_Q$, then $\alpha(f_1(r)) = \alpha(r)$.

(2) If $a_1, a_2, a_3, \dots, a_n \in \mathbf{El}_Q$, $r \in \mathbf{Rel}^1_Q$ and $\alpha(r) = n$, then

$$(a_1, a_2, a_3, \dots, a_n) \in r \Rightarrow (f_0(a_1), f_0(a_2), \dots, f_0(a_n)) \in f_1(r)$$

For all $m = 2, 3, \dots, k$, we have f_m: $\mathbf{Rel}^m_Q \to \mathbf{Rel}^m_R$ is a graded mapping, i.e., a mapping that preserves the arity of relations, such that it satisfies the following conditions:

(1m) If $r \in \boldsymbol{Rel}^m_Q$, then $\alpha(f_m(r)) = \alpha(r)$.

(2m) If $r \in \boldsymbol{Rel}^m_Q$, $A_i \in \{ \mathbf{El}_Q, \mathbf{Rel}^1_Q, \mathbf{Rel}^2_Q, \dots, \mathbf{Rel}^{m-1}_Q \}$, $b_i \in A_i$ $(i = 1, 2, 3, \dots, n)$, $r \subseteq A_1 \times A_2 \times \dots \times A_n$ and $\alpha(r) = n$, then

$$(b_1, b_2, b_3, \dots, b_n) \in r \Rightarrow (f_{i_1}(b_1), f_{i_2}(b_2), \dots, f_{i_n}(b_n)) \in f_m(r)$$

where f_{i_j}: $A_j \to B_j$, $B_j \in$ { $\mathbf{El}_R$, $\mathbf{Rel}^1{}_R$, $\mathbf{Rel}^2{}_R$, … , $\mathbf{Rel}^{m-1}{}_R$ }, A_j and B_j have the same level, and $f_{i_j} = f_t \in F$ for a relevant index t (j = 1, 2, 3, … , n).

We see that a structural homomorphism F: $Q \to R$ of the level k is a graded mapping that respects (preserves) relations in the structure Q which have the level k and less.

Example C16. Homorphisms of groups are structural homomorphisms of the level 2.

Example C17. Linear mappings of linear spaces are structural homomorphisms of the level 2.

Proposition C2. If Q is a structure of the level k and F: $Q \to R$ is a structural homomorphism of the level k, then F is a structural homomorphism of the level t whenever t is larger than k.

Indeed, as Q does not have relations that have the level larger than k, all conditions from Definition C23 for relations that have the level larger than k are satisfied automatically.

Proposition C3. If F: $Q \to R$ is a structural homomorphism of the level k, then F is a structural homomorphism of the level t for any t that is less than k.

Proof follows directly from Definition C24.

An important operation with structural homomorphisms of the level k is the sequential composition of structural homomorphisms of the level k. It is defined by the sequential composition of the coordinates.

Theorem C1. The sequential composition of structural homomorphisms of the level k is a structural homomorphism of the level k.

Proof. Let us consider three set-theoretical structures Q, R and P with two structural homomorphisms F: $Q \to R$ and G: $R \to P$ of the level k. It means that there are two sequences $F = (f_0, f_1, \dots, f_k)$ and $G = (g_0, g_1, \dots, g_k)$ of graded mappings f_0: $\mathbf{El}_Q \to \mathbf{El}_R$, g_0: $\mathbf{El}_R \to \mathbf{El}_P$, f_m: $\mathbf{Rel}^m{}_Q \to \mathbf{Rel}^m{}_R$ and g_m: $\mathbf{Rel}^m{}_R \to \mathbf{Rel}^m{}_P$ (m = 1, 2, 3, … , k). According to the traditional approach, we define the sequential composition $G \circ F$ of F and G as the sequence $(g_0 \circ f_0, g_1 \circ f_1, \dots, g_k \circ f_k)$. To show that $G \circ F$ is a structural homomorphism of the level k, we have to check conditions from Definition C24. It is done by induction. At first, we prove the statement for the first two levels and then for the general case.

(1) By the definition of the sequential composition of functions (see Appendix A), $g_0 \circ f_0$ is a mapping from $\mathbf{El}_Q$ into $\mathbf{El}_P$.

(2) By the definition of the sequential composition of function (see Appendix A), $g_1 \circ f_1$ is a mapping from $\mathbf{Rel}^1{}_Q$ into $\mathbf{Rel}^1{}_P$. Besides, if $r \in \mathbf{Rel}^1{}_Q$, then $\alpha(g_1(f_1(r))) = \alpha(f_1(r)) = \alpha(r)$ as both F and G are structural homomorphisms. In addition, if $a_1, a_2, a_3, \dots, a_n \in \mathbf{El}_Q$, $r \in \mathbf{Rel}^1{}_Q$ and $\alpha(r) = n$, then

$$(a_1, a_2, a_3, \dots, a_n) \in r \Rightarrow (f_0(a_1), f_0(a_2), \dots, f_0(a_n)) \in f_1(r)$$

and

$$(f_0(a_1), f_0(a_2), \dots, f_0(a_n)) \in f_1(r) \Rightarrow (g_0(f_0(a_1)), g_0(f_0(a_2)), \dots, g_0(f_0(a_n))) \in g_1(f_1(r))$$

Thus,

$$(a_1, a_2, a_3, \ldots, a_n) \in r \Rightarrow (g_0(f_0(a_1)), g_0(f_0(a_2)), \ldots, g_0(f_0(a_n))) \in g_1(f_1(r))$$

(3) For all $m = 2, 3, \ldots, k$, the composition $g_m \circ f_m$ is a graded mapping $g_m \circ f_m$: $\mathbf{Rel}^m{}_Q \to \mathbf{Rel}^m{}_P$ by the definition of the sequential composition of function (see Appendix A). Besides, if $r \in \boldsymbol{Rel}^m{}_Q$, then $\alpha(g_m(f_m(r)) = \alpha(f_m(r)) = \alpha(r)$ as both F and G are structural homomorphisms. In addition, if $A_i \in \{ \mathbf{El}_Q, \mathbf{Rel}^1{}_Q, \mathbf{Rel}^2{}_Q, \ldots, \mathbf{Rel}^{m-1}{}_Q \}$, $b_i \in A_i$ ($i = 1, 2, 3, \ldots, n$), $r \subseteq A_1 \times A_2 \times \ldots \times A_n$, $r \in \boldsymbol{Rel}^m{}_Q$ and $\alpha(r) = n$, then

$$(b_1, b_2, b_3, \ldots, b_n) \in r \Rightarrow (f_{i_1}(b_1), f_{i_2}(b_2), \ldots, f_{i_n}(b_n)) \in f_m(r)$$

and

$$(f_{i_1}(b_1), f_{i_2}(b_2), \ldots, f_{i_n}(b_n)) \in f_m(r) \Rightarrow (g_{i_1}(f_{i_1}(a_1)), g_{i_2}(f_{i_2}(a_2)), \ldots, g_{i_n}(f_{i_n}(a_n))) \in g_m(f_m(r))$$

Thus,

$$(b_1, b_2, b_3, \ldots, b_n) \in r \Rightarrow (g_{i_1}(f_{i_1}(a_1)), g_{i_2}(f_{i_2}(a_2)), \ldots, g_{i_n}(f_{i_n}(a_n))) \in g_m(f_m(r))$$

where f_{i_j}: $A_j \to B_j$, g_{i_j}: $B_j \to C_j$, $B_j \in \{ \mathbf{El}_R, \mathbf{Rel}^1{}_R, \mathbf{Rel}^2{}_R, \ldots, \mathbf{Rel}^{m-1}{}_R \}$, $C_j \in \{ \mathbf{El}_P, \mathbf{Rel}^1{}_P, \mathbf{Rel}^2{}_P, \ldots, \mathbf{Rel}^{m-1}{}_P \}$, A_j, B_j and C_j have the same level, $f_{i_j} = f_t$ and $g_{i_j} = g_t$ for a relevant index t.

Theorem is proved.

Definition C25. A *structural homomorphism* F: $Q \to R$ is a sequence (graded mapping) $F = (f_0, f_1, f_2, f_3, \ldots, f_m, \ldots)$ such that the beginning (head) $(f_0, f_1, \ldots, f_k)$ of F is a structural homomorphism of the level k for all $k = 1, 2, 3, \ldots$.

Example C18. Homorphisms of groups are structural homomorphisms (Kurosh, 1963).

Example C19. Linear mappings of linear spaces are structural homomorphisms (Kurosh, 1963).

Example C20. Homorphisms of universal algebras are structural homomorphisms (Cohn, 1965).

Example C21. Open mappings of topological spaces are structural homomorphisms (Kuratowski, 1966).

Example C22. Closed mappings of topological spaces are structural homomorphisms (Kuratowski, 1966).

Example C23. A mapping f: $X \to Y$ of a topological space X into a topological space X is called sequentially continuous if it satisfies the following condition

If $a = \lim_{i=1}{}^{\infty}\{ a_i ; a_i \in X, i = 1, 2, 3, \ldots\}$, then $f(a) = \lim_{i=1}{}^{\infty}\{ f(a_i) ; i = 1, 2, 3, \ldots\}$.

Let us take the relation LIM between elements of topological spaces and sequences in these spaces, namely,

$$(a, \{ a_i ; i=1, 2, 3, \ldots\}) \in \text{LIM if } a = \lim_{i=1}{}^{\infty} a_i$$

Then structural homomorphisms with respect to the relation LIM coincide with sequentially continuous mappings of topological spaces.

Propositions C2 and C3 imply the following result.

Corollary C1. If Q is a structure of the level k and F: $Q \to R$ is a structural homomorphism of the level k, then F is a structural homomorphism.

Theorem C1 implies the following result.

Corollary C2. The sequential composition of structural homomorphisms is a structural homomorphism.

This result shows that it is possible to consider a category **Str** in which objects are structures and morphisms are structural homomorphisms.

Definition C26. a) A structural homomorphism $F = (f_0 , f_1 , \ldots , f_k)$: $Q \to R$ of the level k is called *saturated* if all $f_1 , \ldots , f_k$ are one-to-one mappings (bijections).

b) A structural homomorphism $F = (f_0 , f_1 , \ldots , f_k)$: $Q \to R$ is called *saturated* if all f_k are one-to-one mappings ($k = 1, 2, 3, \ldots$).

Proposition C4. If Q and R are structures of the level k and F: $Q \to R$ is a saturated structural homomorphism of the level k, then F is a structural homomorphism of the level t whenever t is larger than k.

Indeed, because the structure Q does not have relations of the levels higher than k, all conditions from Definition C25 are automatically satisfied.

Proposition C5. If F: $Q \to R$ is a saturated structural homomorphism of the level k, then F is a saturated structural homomorphism of the level t for any t that is less than k.

Proof follows directly from Definition C26.

Corollary C3. If Q and R are structures of the level k and F: $Q \to R$ is a saturated structural homomorphism of the level k, then F is a saturated structural homomorphism.

This result follows directly from Propositions C4 and C5.

Theorem C2. The sequential composition of saturated structural homomorphisms of the level k is a saturated structural homomorphism of the level k.

Indeed, by Theorem C1, the sequential composition of structural homomorphisms of the level k is a structural homomorphism of the level k and the composition of one-to-one mappings is a one-to-one mapping.

Theorem C2 implies the following result.

Corollary C4. The sequential composition of saturated structural homomorphisms is a saturated structural homomorphism.

This result shows that it is possible to consider a category **SStr** in which objects are structures and morphisms are saturated structural homomorphisms. The category **SStr** is a subcategory of the category **Str**.

Let us consider two structures Q and R.

Definition C27. A structure Q (of the level $q \geq 0$) is a *substructure* of the the level $k \geq 0$ of a structure R (of the level $r \geq 0$), it is denoted by $Q \subseteq^k R$, if:

(a) if $k = 0$, then $\mathbf{El}_Q \subseteq \mathbf{El}_R$;

(b) if $k > 0$, then $\mathbf{El}_Q \subseteq \mathbf{El}_R$ and all relations of the level k and less in Q are induced by relations of the level k and less in R.

It means that there is a sequence $(\mu_0 , \mu_1 , \ldots , \mu_k)$ such that:

(1) μ_0 is the standard inclusion of $\mathbf{El}_Q \subseteq \mathbf{El}_R$, that is, $\mu_0(a) = a$ for any $a \in \mathbf{El}_Q$.
(2) μ_1: $\mathbf{Rel}^1_Q \to \mathbf{Rel}^1_R$ is a graded injection such that

$$\forall q \in \mathbf{Rel}^1_Q \text{ with } \alpha(q) = n \ (q = \mu_1(q) \cap (\mathbf{El}_Q)^n)$$

(3) for all $t = 2, \dots k$, μ_m: $\mathbf{Rel}^m_Q \to \mathbf{Rel}^m_R$ is a graded injection, i.e., $\alpha(\mu_m(q)) = \alpha(q)$ for any $q \in \mathbf{Rel}^t_Q$, such that

If $A_i \in \{ \mathbf{El}_Q, \mathbf{Rel}^1_Q, \mathbf{Rel}^2_Q, \dots, \mathbf{Rel}^{m-1}_Q \}$, $b_i \in A_i$ $(i = 1, 2, 3, \dots, n)$, $\alpha(b_i) = n_i$, $r \subseteq A_1 \times A_2 \times \dots \times A_n$, $r \in \mathbf{Rel}^m_Q$, $\alpha(r) = n$ and $(b_1, b_2, b_3, \dots, b_n) \in r$, then

$$r = \mu_m(r) \cap (A_1 \times A_2 \times \dots \times A_n)$$

Note that in the set-theoretical model, a substructure of the zero level is simply a subset of the set of all elements of R.

Example C24. A subgroup H of a group G is a substructure of the second level of G.

Example C25. A subspace M of a linear space L is a substructure of the second level of L.

These examples are ordinary constructions in algebra. However, it is also possible to show that some constructions that are informally used in mathematics are also encompassed by the concept of a substructure.

Example C26. A subsemigroup H of a group G, e.g., $N \subseteq \boldsymbol{Z}$ or $N \subseteq \boldsymbol{R}$, is a substructure of the level 2 of G.

Example C27. The additive subgroup M of a linear space L is a substructure of the level 2 of L.

Proposition C6. If a structure Q is a substructure of the level k of a structure R, then for any $t < k$, Q is a substructure of the level t of R.

Proof follows directly from Definition C27.

Proposition C7. If a structure Q of the level k is a substructure of the level k of a structure R, then Q is a substructure of R of any level t such that $t > k$.

Indeed, as Q does not have relations that have the level larger than k, all conditions from Definition C27 for relations that have the level larger than k are satisfied automatically.

Proposition C8. If a structure Q is a substructure of the level k of a structure R and the structure R is a substructure of the level k of a structure P, then Q is a substructure of the level k of P.

Proof is similar to the proof of Theorem C1 taking into account that the sequential composition of injections is an injection and the sequential composition of inclusions is an inclusion.

Proposition C8 means that the relation "to be a substructure of the level k" is transitive.

Definition C28. A structure Q is a *substructure* of a structure R if Q is a substructure of all levels $k = 0, 1, 2, 3, \dots$ of R. It is denoted by $Q \subseteq R$.

Example C28. A subgroup H of a group G is a substructure of G.

Example C29. A subspace M of a linear space L is a substructure of L.

Example C30. A subsemigroup H of a group G, e.g., $N \subseteq \boldsymbol{Z}$ or $N \subseteq \boldsymbol{R}$, is a substructure of G.

Example C31. Any additive subgroup M of a linear space L is a substructure of L.

Proposition C9. If a structure Q is a substructure of a structure R of the level k, then the level of the structure Q cannot be larger than k.

Indeed, if $t > k$ and $\mathbf{Rel}^k{}_Q \neq \varnothing$, then by Definition C27, there is an injection μ_t: $\mathbf{Rel}^t{}_Q \to \mathbf{Rel}^t{}_R$. However, $\mathbf{Rel}^k{}_R = \varnothing$,. Thus, $\mathbf{Rel}^k{}_Q = \varnothing$, for all $t > k$.

Proposition C10. Any structure Q of the level k that is a substructure of the level k of a structure R is a substructure of R.

Proof. If a structure Q of the level k that is a substructure of the level k of a structure R, then the structure Q does not have relations of the levels higher than k, all conditions from Definition C27 are automatically satisfied.

Proposition C8 implies the following result.

Corollary C5. If a structure Q is a substructure of a structure R and the structure R is a substructure of a structure P, then Q is a substructure of P.

Corollary C4 means that the relation "to be a substructure" is transitive.

Definition C29. a) A substructure Q of the level k of a structure R that is defined by a sequence $(\mu_0 , \mu_1 , \dots , \mu_k)$ is called *saturated* if all $\mu_1 , \dots , \mu_k$ are one-to-one mappings. It is denoted by $Q \subseteq^k{}_s R$.

b) A substructure Q of a structure R that is defined by a sequence $(\mu_0 , \mu_1 , \dots , \mu_k , \dots)$ is called *saturated* if all μ_k are one-to-one mappings ($k = 1, 2, 3, \dots$). It is denoted by $Q \subseteq_s R$.

Example C32. A subgroup H of a group G is a saturated substructure of G.

Example C33. A subspace M of a linear space L is a saturated substructure of L.

Example C34. A subsemigroup H of a group G is a substructure of G, but it is not a saturated substructure of G if H is not a subgroup of G. For instance, $\boldsymbol{N}$ is the semigroup with respect to addition of all natural numbers, then $\boldsymbol{N}$ is a substructure of the group $\boldsymbol{Z}$ of all integer numbers but it is not a saturated substructure of $\boldsymbol{Z}$.

Proposition C11. If a structure Q is a saturated substructure of the level k of a structure R and the structure R is a saturated substructure of the level k of a structure P, then Q is a saturated substructure of the level k of P.

Proposition C11 means that the relation "to be a saturated substructure of the level k" is transitive.

Corollary C6. If a structure Q is a saturated substructure of a structure R and the structure R is a saturated substructure of a structure P, then Q is a saturated substructure of P.

Corollary C6 means that the relation "to be a saturated substructure" is transitive.

Proposition C12. If a structure Q is a saturated substructure of the level k of a structure R, then for any $t < k$, Q is a saturated substructure of the level t of R.

Proof follows directly from Definitions C27 and C29.

Definition C30. A structure Q (of the level $q \geq 0$) is a *weak substructure* of the level $k \geq 0$ of a structure R (of the level $r \geq 0$), it is denoted by $Q \subseteq^k{}_w R$, if:

(a) if $k = 0$, then $\mathbf{El}_Q \subseteq \mathbf{El}_R$;

(b) if $k > 0$, then $\mathbf{El}_Q \subseteq \mathbf{El}_R$ and there is a sequence $(\mu_0 , \mu_1 , \dots , \mu_k)$ such that:

(1) μ_0 is the standard inclusion of $\mathbf{El}_Q \subseteq \mathbf{El}_R$, that is, $\mu_0(a) = a$ for any $a \in \mathbf{El}_Q$.

(2) μ_1: $\mathbf{Rel}^1{}_Q \to \mathbf{Rel}^1{}_R$ is a graded injection such that

$$\forall q \in \mathbf{Rel}^1{}_Q \text{ with } \alpha(q) = n \ (q \subseteq \mu_1(q) \cap (\mathbf{El}_Q)^n)$$

(3) for all $t = 2, \ldots k$, μ_t: $\mathbf{Rel}^t_Q \to \mathbf{Rel}^t_R$ is a graded injection, i.e., $\alpha(\mu_t(q)) = \alpha(q)$ for any $q \in \mathbf{Rel}^t_Q$, such that

If $A_i \in \{ \mathbf{El}_Q , \mathbf{Rel}^1_Q , \mathbf{Rel}^2_Q , \ldots , \mathbf{Rel}^{m-1}_Q \}$, $b_i \in A_i$ ($i = 1, 2, 3, \ldots , n$), $\alpha(b_i) = n_i$, $r \subseteq A_1 \times A_2 \times \ldots \times A_n$, $r \in \boldsymbol{Rel}^m_Q$, $\alpha(r) = n$ and $(b_1 , b_2 , b_3 , \ldots , b_n) \in r$, then

$$r \subseteq \mu_t(r) \cap (A_1 \times A_2 \times \ldots \times A_n)$$

Note that in the set-theoretical model, a weak substructure of the zero level is simply a subset of the set of all elements of *R*.

Example C35. Let us consider the set *A* of all natural numbers with addition defined only for even numbers and the semigroup with respect to addition ***N*** of all natural numbers. Then *A* is a weak substructure of the level 2 of ***N*** but it is not a substructure of the level 2 of ***N***.

Definitions imply the following result.

Proposition C13. Any substructure of a structure *R* is a weak substructure of *R*.

It means that the relation "to be a substructure" is stronger than the relation "to be a weak substructure".

Proposition C14. If a structure *Q* is a weak substructure of the level *k* of a structure *R*, then for any $t < k$, *Q* is a weak substructure of the level *t* of *R*.

Proof follows directly from Definition C30.

Proposition C15. If a structure *Q* of the level *k* is a weak substructure of the level *k* of a structure *R*, then for any $t > k$, *Q* is a weak substructure of the level *t* of *R*.

Proof is similar to the proof of Proposition C7.

Proposition C16. If a structure *Q* is a weak substructure of the level *k* of a structure *R* and the structure *R* is a weak substructure of the level *k* of a structure *P*, then *Q* is a weak substructure of the level *k* of *P*.

Proof is similar to the proof of Proposition C8.

Proposition C16 means that the relation "to be a weak substructure of level *k*" is transitive.

Definition C31. A structure *Q* is a *weak substructure* of a structure *R* if *Q* is a weak substructure of all levels $k = 0, 1, 2, 3, \ldots$ of *R*. It is denoted by $Q \subseteq_{\text{w}} R$.

Example C36. Let us consider the set *A* of all natural numbers with addition defined only for even numbers and the semigroup with respect to addition ***N*** of all natural numbers. Then *A* is a weak substructure of ***N*** but it is not a substructure of ***N***.

Proposition C17. If a structure *Q* is a weak substructure of a structure *R* of the level *k*, then the level of the structure *Q* cannot be larger than *k*.

Indeed, if $t > k$ and $\mathbf{Rel}^k_Q \neq \varnothing$, then by Definition C30, there is an injection μ_t: $\mathbf{Rel}^t_Q \to \mathbf{Rel}^t_R$. However, $\mathbf{Rel}^k_R = \varnothing$,. Thus, $\mathbf{Rel}^k_Q = \varnothing$, for all $t > k$.

Proposition C16 implies the following result.

Corollary C7. If a structure *Q* is a weak substructure of a structure *R* and the structure *R* is a weak substructure of a structure *P*, then *Q* is a weak substructure of *P*.

Corollary C4 means that the relation "to be a weak substructure" is transitive.

Definition C32. A structure *R* is an *enrichment* of a structure *Q* if $\mathbf{El}_Q = \mathbf{El}_R$ and *Q* is a substructure of *R*.

Proposition C18. If a structure *Q* is an enrichment of a structure *R* and the structure *R* is an enrichment of a structure *P*, then *Q* is an enrichment of *P*.

Indeed, by Definition C32, $\mathbf{El}_P = \mathbf{El}_R = \mathbf{El}_Q$ and by Proposition C8, P is a substructure of Q. So, by Definition C32, Q is an enrichment of P.

Proposition C19. If a structure P is a substructure of the level 0 of a structure Q, then there is an enrichment R of a structure Q such that P is a substructure of R.

Proof. As the structure P is a substructure of the level 0 of the structure Q, by Definition C27, $\mathbf{El}_P \subseteq \mathbf{El}_Q$. Note that by Definition C32, $\mathbf{El}_R = \mathbf{El}_Q$. Thus, to any relation $q \in \mathbf{Rel}^1_P$ with $\alpha(q) = n$, it is possible to correspond an n-ary relation r in the set $\mathbf{El}_Q$, such that

$$\forall\, a_1, a_2, a_3, \ldots, a_n \in \mathbf{El}_Q\, ((a_1, a_2, a_3, \ldots, a_n) \in r \Leftrightarrow (\, a_1, a_2, a_3, \ldots , a_n \in \mathbf{El}_P \,\&\, (a_1, a_2, a_3, \ldots, a_n) \in q)$$

Then we assume that all such relations r belong to the structure R that we are building, i.e., $\mathbf{Rel}^1_R = \mathbf{Rel}^1_Q \cup \mathbf{Rel}^1_{\text{new}}$, where $\mathbf{Rel}^1_{\text{new}}$ is the set of all new relations r. As q is an arbitrary relation from $\mathbf{Rel}^1_P$, in such a way, we obtain a mapping μ_1: $\mathbf{Rel}^1_P \to \mathbf{Rel}^1_R$, which is a graded injection such that

$$\forall q \in \mathbf{Rel}^1_P \text{ with } \alpha(q) = n\ (q = \mu_1(q) \cap (\mathbf{El}_R)^n\,)$$

In similar a way, we correspond relations of the higher levels in the new structure R to the relations of the higher levels from the structure P and add these new relations to the relations of the higher levels from the structure Q. By construction, this allows us to satisfy conditions from Definition 27, affirming that P is a substructure of R. Besides, as all relations from Q also belong to R, the structure R is an enrichment of the structure Q.

Proposition is proved.

Definition C33. a) The *image* ImF of a structural homomorphism $F = (f_0 , f_1 , \ldots , f_m , \ldots)$: $Q \to R$ of the level k is the structure defined by the images of all mappings $f_0 , f_1 , f_2 , f_3 , \ldots , f_k$, that is, $\mathbf{El}_{\text{Im}F} = \text{Im}\, f_0$, $\mathbf{Rel}^1_{\text{Im}F} = \text{Im}\, f_1$, $\mathbf{Rel}^2_{\text{Im}F} = \text{Im}\, f_2$, … , $\mathbf{Rel}^k_{\text{Im}F} = \text{Im}\, f_k$.

b) The *image* ImF of a structural homomorphism $F = (f_0 , f_1 , \ldots , f_m , \ldots)$: $Q \to R$ is the structure defined by the images of all mappings $f_0 , f_1 , f_2 , f_3 , \ldots , f_m , \ldots$, that is, $\mathbf{El}_{\text{Im}F} = \text{Im}\, f_0$, $\mathbf{Rel}^1_{\text{Im}F} = \text{Im}\, f_1$, $\mathbf{Rel}^2_{\text{Im}F} = \text{Im}\, f_2$, … , $\mathbf{Rel}^m_{\text{Im}F} = \text{Im}\, f_m$, …

Images of homorphisms of groups, rings and other universal algebras are examples of images of structural homorphisms.

Theorem C3. The image of a structural homomorphism F: $Q \to R$ of the level n is a weak substructure of the level n of R.

Proof. Let us consider two set-theoretical structures Q and R with a structural homomorphisms F: $Q \to R$ of the level k. It means that there is a sequences $F = (f_0 , f_1 , \ldots , f_k)$ of graded mappings f_0: $\mathbf{El}_Q \to \mathbf{El}_R$ and f_m: $\mathbf{Rel}^m_Q \to \mathbf{Rel}^m_R$ ($m = 1, 2, 3, \ldots , k$). By Definition C33, the image ImF of a structural homomorphism F is the structure in which $\mathbf{El}_{\text{Im}F} = \text{Im}\, f_0$, $\mathbf{Rel}^1_{\text{Im}F} = \text{Im}\, f_1$, $\mathbf{Rel}^2_{\text{Im}F} = \text{Im}\, f_2$, … , $\mathbf{Rel}^k_{\text{Im}F} = \text{Im}\, f_k$.

To show that ImF is a weak substructure of the level n of R, we have to check conditions from Definition C30.

(1) By the definition of the image of a function (see Appendix A), $\mathbf{El}_{\text{Im}F} \subseteq \mathbf{El}_R$ and by the definition of a structural homomorphism, there is the standard inclusion μ_0 of $\mathbf{El}_Q$ into $\mathbf{El}_R$.

(2) By the same argument, there is a graded injection μ_1: $\mathbf{Rel}^1_{\text{Im}F} \to \mathbf{Rel}^1_{\text{R}}$. Let us consider elements for $b_1 , b_2 , b_3 , \ldots , b_n \in \mathbf{El}_{\text{Im}F}$ and $q \in \mathbf{Rel}^1_{\text{Im}F}$ with $\alpha(q) = n$. By the definition of the image of a function, there are $a_1 , a_2 , a_3 , \ldots , a_n \in \mathbf{El}_Q$ and $r \in \mathbf{Rel}^1_Q$ such

that $b_i = f_0(a_i)$ (= 1, 2, 3, … , n) and $q = f_1(r)$. By the definition of a structural homomorphism, for $a_1, a_2, a_3, \dots, a_n \in \mathbf{El}_Q$ and $q \in \mathbf{Rel}^1_Q$ with $\alpha(q) = n$, we have

$$(a_1, a_2, a_3, \dots, a_n) \in r \Rightarrow (f_0(a_1), f_0(a_2), \dots, f_0(a_n)) \in f_1(r)$$

This implies that

$$(b_1, b_2, b_3, \dots, b_n) \in q \Rightarrow (\mu_0(b_1), \mu_0(b_2), \dots, \mu_0(b_n)) \in \mu_1(r)$$

Thus, we have

$$q \subseteq \mu_1(q) \cap (\mathbf{El}_{\mathrm{Im}F})^n$$

(3) In a similar way, for all $t = 2, \dots k$, we define graded injections μ_t: $\mathbf{Rel}^t_Q \to \mathbf{Rel}^t_R$ for which the general condition from Definition C30 is also satisfied.

Theorem is proved.

Corollary C8. The image of a structural homomorphism f: $Q \to R$ is a weak substructure of R.

Definition C34. A structural homomorphism f: $Q \to R$ is called:

a) a *structural monomorphism of the level n* if all mappings $f_0, f_1, f_2, f_3, \dots, f_n$ are injections;

b) a *structural monomorphism on the level n* if the mapping f_n is an injection.

c) a *structural monomorphism* if f is a structural monomorphism of all levels k = 1, 2, 3, …

Theorem C4. The sequential composition of structural monomorphisms of the level n (on the level n) is a structural monomorphism of the level n (on the level n).

Proof is similar to the proof of Theorem C1 and is based on properties of injections of sets.

Corollary C9. The sequential composition of structural monomorphisms is a structural monomorphism.

Definition C35. A structural homomorphism f: $Q \to R$ is called:

a) a *structural epimorphism of the level n* if all mappings $f_0, f_1, f_2, f_3, \dots, f_n$ are projections;

b) a *structural epimorphism on the level n* if the mapping f_n is a projection.

c) a *structural epimorphism* if f is a structural epimorphism of all levels k = 1, 2, 3, …

There are structural homomorphisms that are structural monomorphisms on some levels and structural epimorphisms on other levels.

Example C36. The natural structural homomorphism of the semigroup $\mathbf{Z}_+$ with respect to addition of all integer numbers, into the group $\mathbf{Z}$ of all integer numbers is a structural epimorphism on the level 0 and structural monomorphism on other levels.

Theorem C5. The sequential composition of structural epimorphisms of the level n (on the level n) is a structural epimorphism of the level n (on the level n).

Proof is similar to the proof of Theorem C1 and is based on properties of projections (surjections) of sets.

Corollary C10. The sequential composition of structural epimorphisms is a structural epimorphism.

Theorem C6. It is possible to represent any structural homomorphism of the level n as the sequential composition of a structural monomorphism of the level n and a structural epimorphism of the level n.

Proof. Let us consider two set-theoretical structures Q and R with a structural homomorphism $F: Q \to R$ of the level n. It means that there is a sequence $F = (f_0, f_1, \dots, f_n)$ of graded mappings f_0: $\mathbf{El}_Q \to \mathbf{El}_R$ and f_m: $\mathbf{Rel}^m{}_Q \to \mathbf{Rel}^m{}_R$ ($m = 1, 2, 3, \dots, n$). By the properties of set mappings, we have that $f_m = g_m \circ p_m$ where p_m is a surjection and g_m is a injection ($m = 0, 1, 2, 3, \dots, n$).

As in the proof of Theorem C3, we check that all mappings p_m and g_m satisfy conditions from Definition C24. This implies that each p_m is a structural epimorphism of the level n and each g_m is a structural monomorphism of the level n ($m = 0, 1, 2, 3, \dots, n$).

Theorem is proved.

Corollary C11. It is possible to represent any structural homomorphism as the sequential composition of a structural monomorphism and a structural epimorphism.

Many properties of algebraic systems and topological spaces are corollaries of the results proved here for general structures. For instance, any homorphism of groups or of rings is the sequential composition of a group (ring) monomorphism and a group (ring) epimorphism (Kurosh, 1963).

Definition C36. A structural homomorphism $f: Q \to R$ is called:

a) a *structural isomorphism of the level n* if it is both a structural epimorphism of the level n and a structural monomorphism of the level n;

b) a *structural isomorphism on the level n* if it is both a structural epimorphism on the level n and a structural monomorphism on the level n.

c) a *structural isomorphism* if f is a structural isomorphism of all levels $k = 1, 2, 3, \dots$

Definitions imply the following results.

Proposition C20. a) If $F = (f_0, f_1, f_2, f_3, \dots)$ is a structural isomorphism, then all mappings $f_0, f_1, f_2, f_3, \dots$ are bijections.

b) If $F = (f_0, f_1, f_2, f_3, \dots, f_n)$ is a structural isomorphism of the level n, then all mappings $f_0, f_1, f_2, f_3, \dots, f_n$ are bijections.

Proposition C21. a) Any structural isomorphism is a saturated structural homomorphism.

b) Any structural isomorphism of the level n is a saturated structural homomorphism of the level n.

Theorems C4 and C5 imply the following result.

Theorem C7. The sequential composition of structural isomorphisms of the level n is a structural isomorphism of the level n.

Corollaries C9 and C10 imply the following result.

Corollary C12. The sequential composition of structural isomorphisms is a structural isomorphism.

To conclude, we would like to emphasize that here we laid the foundations for the general theory (metatheory) of structures in the set-theoretical context making only first steps in its development. At the same time, we can see that this is a big area in mathematics, which waits for new researchers to explore it.

There are many directions for research in this area. For instance, it is possible to develop a similar theory and obtain corresponding results for mereological structures taking parts instead of elements. Here we only show that a subset of a set is substructure of this set in the sense of Definitions C26 and C27 interpreted for mereological structures.

Let us consider a set B and its subset A. A set is defined by the membership relation $\in$, i.e., it is determined when $x \in B$. Specializing this relation for a definite set, we denote by $\in_X$ the membership relation in a set X and take all elements from the set and the set itself as its parts. To define A as a substructure of B, we need injections μ_0 and μ_1. They are defined in the following way: $\mu_0(a) = a$, $\mu_0(A) = B$, and $\mu_0(\in_A) = \in_B$. Then by the definition of a subset, we have

$$\in_A = \in_B \cap (\mathbf{Parts}_A)^2 = \mu_0(\in_A) \cap (\mathbf{Parts}_A)^2$$

This means that A is a substructure of B. because there is a natural injection of all parts $\mathbf{Parts}_A$ of A into all parts $\mathbf{Parts}_B$ of B.

REFERENCES

Abadi, M. and Cardelli, L. *A Theory of Objects*, Springer-Verlag, New York, 1996.

Aizerman, M. A., Gusev, L. A., Petrov, S. V., Smirnova, I. M. and Tenenbaum, L. A. Dynamic approach to analysis of structures described by graphs (foundations of graph-dynamics), in *Topics in the General Theory of Structures*, D. Reidel Publishing Company, Dordrecht, Holland, 1987.

Aristotle, *The Complete Works of Aristotle*, Princeton University Press, Princeton, 1984.

Armstrong, S.L., Gleitman, L. R. and Gleitman, H. (1983) What some concepts might not be, *Cognition*, v. 13, pp. 263-308.

Artaud, A. *Le Théâtre et son Double*, Gallimard, Paris, 1938.

Audi, R. *The Structure of Justification*, Cambridge University Press, New York, 1993.

Awodey, S. (1996) Structuralism in Mathematics and Logic: A Categorical Perspective, *Philosophia Mathematica*, v. 4, pp. 209–37.

Awodey, S. (2004) An Answer to G. Hellman's Question "Does Category Theory Provide a Framework for Mathematical Structuralism?" *Philosophia Mathematica*, v. 12, pp. 54–6.

Baç, M. (2000) Structure versus Process: Mach, Hertz, and the Normative Aspect of Science, *Journal for General Philosophy of Science*, Zeitschrift für allgemeine Wissenschaftstheorie, v. 31, pp. 19-37.

Baghirathan, R., Rada, C. and Taylor, L. (2004) Structuralist economics: worldly philosophers, models, and methodology, *Social Research*, v. 71, No. 2, pp. 305-326.

Balcazar, J.L., Diaz, J., and Gabarro, J. *Structural Complexity*, Springer-Verlag, Berlin/Heidelberg/ New York, 1988.

Balzer, W., Burgin, M. and Kuznetsov, V. (1991) Reduction and the Structure-Nominative View of Theories, in Abstracts of the 9th *International Congress of Logic, Methodology and Philosophy of Science*, Uppsala, Sweden, v. II, p. 6.

Balzer, W., Moulines, C.U. and Sneed, J.D. *An Architectonic for Science. The Structuralist Program*, Reidel, Dordrecht, 1987.

Banerjee, D.K ., Parui, S.K. and Majumder, D.D. (1990) Shape based object recognition, *Lecture Notes in Computer Science*, v. 444, pp. 363-374.

Barantsev, R.G. (1982) System Triad of Definition, *International forum on information and documentation*, v. 7, No. 1, pp. 9-13.

Barantsev, R.G. (1989) The system triad as the structural cell of the synthesis, System researches. Methodological problems, Nauka, Moscow, pp. 193 - 209 (in Russian).

Barfield, W. and Caudell, T. (Eds.) *Fundamentals of Wearable Computers and Augmented Reality*, Lawrence Erlbaum, Mahwah, NJ, 2001.

Bastide, R. (Ed.) *Sens et usages du terme Structure dans les sciences humaines et sociales*, s'-Gravenhage, 1962.

Bell, J.L. *Set Theory: Boolean-Valued Models and Independence Proofs*, Clarendon Press, Oxford, 2005.

Berge, C. *Graphs and Hypergraphs*, North Holland P.C., Amsterdam/New York, 1973.

Berger, P. L. and Luckmann, T. *The Social Construction of Reality: A Treatise in the Sociology of Knowledge*, Anchor Books, Garden City, NY, 1966.

Bernays, P. (1935) Sur le platonisme dans les mathematiques, *L'Enseignement math.*, v. 34, pp. 52-69.

Bernays, P. (1967) Hilbert, David, in *The encyclopedia of philosophy*, v. 3, Macmillan P.C. and The Free Press, New York, pp. 496-504.

Berne, E. *Games People Play*, Grove Press, New York, 1964.

Beth, E.W. and Piaget, J. *Mathematical Epistemology and Psychology*, Springer, New York, 1974.

Biederman, I. (1987) Recognition by components: A theory of human image understanding, *Psychological Review*, v. 94, No. 2, pp. 115-147.

Biederman, I. and Cooper, E.E. (1992) Size invariance in visual object priming, *J. Exper. Psychology and Human Perception Performance*, v. 18, pp. 121-133.

Bimber, O. and Raskar, R. *Spatial Augmented Reality: Merging Real and Virtual Worlds*, A K Peters, 2005.

Bitar, S. (1988) Neoconservatism versus Neostructuralism in Latin America, *CEPAL Review*, No. 34, pp. 44-62

Bojowald, M, and Skirzewski, A. (2008) Effective theory for the cosmological generation of structure *Advanced Science Letters*, v. 1, pp. 92-98.

Bollen, K. *Structural Equations with Latent Variables*, John Wiley, New York, 1989.

Bononi, A. (Ed.) *La struttura logica del linguaggio*, Milano, 1973.

Born, M. (1953) Physical Reality, *The Philosophical Quarterly*, v. 3, No. 11, pp. 139-149.

Bourbaki, N. (1948) L'architecture des mathématiques, Legrands courants de la pensée mathématiques, *Cahiers Sud*, pp.35-47.

Bourbaki, N. *Structures*, Hermann, Paris, 1957.

Bourbaki, N. *Theorie des Ensembles*, Hermann, Paris, 1960 (English translation: Bourbaki, N. *Theory of Sets*, Hermann, Paris, 1968).

Boyle, R. *The Sceptical Chymist*, (reprint) Dover Publications, Inc. New York, 1661.

Brass, P. *Advanced Data Structures*, Cambridge University Press, Cambridge, 2008..

Brendal V. (1939) Linguistique structurale, *Acta linguistica*, v. 1, No. 1, pp. 2-10.

Brenner, J. E. *Logic in Reality*, Springer, Dordrecht, 2008.

Brooks, C. *The Well Wrought Urn*, Harcourt Brace, New York, 1947.

Brooks, C. Irony as a Principle of Structure, in *Critical Theory Since Plato*, Harcourt Brace Jovanovich, Inc., New York, 1971.

Brown, P. and Lauder, H. (2000) Collective intelligence, in *Social Capital: Critical Perspectives*, Oxford University Press, New York.

Brukner, Č. and Zelinger, A. Information and fundamental elements of the structure of quantum theory, *Festschrift for C. F. von Weizsaecker on the Occasion of his 90th Birthday*, Preprint in Quantum Physics, quant-ph/0212084, 2002 (electronic edition: http://arXiv.org).

Buck, R. C. and Hull, D. L. (1966) The Logical Structure of the Linnean Hierarchy, *Systematic Zoology*, v. 33, pp. 97-111.

Buckley, M. (2010) The structure of justification in political Constructivism, *Metaphilosophy*, v. 41, No. 5, pp. 669-689.

Buckley, M. W. *The Structure of Business*, Pitman Publishing, London, 1990.

Bucur, I. and Deleanu, A. *Introduction to the theory of categories and functors*, John Wiley, London, 1968.

Burgin, M. (1983) On the Greg's paradox in taxonomy, *Abstracts presented to the American Mathematical Society*, v. 4, No. 3, p. 303.

Burgin, M. (1984) Composition of operators in a multidimensional structured model of parallel computations and systems, *Cybernetics and System Analysis*, v. 19, No. 3, pp. 340-350 (translated from Russian).

Burgin, M. (1985) Abstract theory of properties, in *Non-classical Logics*, Institute of Philosophy, Moscow, pp. 109-118 (in Russian).

Burgin, M. *Named Sets, General Theory of Properties, and Logic*, Institute of Philosophy, Kiev, 1989 (in Russian) .

Burgin, M. (1990) Theory of Named Sets as a Foundational Basis for Mathematics, in *Structures in Mathematical Theories*, San Sebastian, pp. 417-420.

Burgin, M. (1990a) Abstract Theory of Properties and Sociological Scaling, in *Expert Evaluation in Sociological Studies*, Kiev, pp. 243-264 (in Russian).

Burgin, M. (1991) Named Sets in the Semantic Network Theory, in *Knowledge - Dialog - Decision*, Leningrad, pp. 43-47 (in Russian).

Burgin, M. (1991b) What is the Surrounding World Built of (The Ontological Aspect), *Philosophical and Sociological Thought*, 1991, No. 8, pp. 54-67 (in Russian).

Burgin, M. (1992) Named sets as a mathematical apparatus for data structure representation, in *Problem solving on mathematical models of object domains*, Kiev, pp. 13-21 (in Russian).

Burgin, M. (1993) Triad as a Fundamental Structure in Human Culture, *Studia Culturologia*, v. 2, pp. 51-63.

Burgin, M. (1993a) On the way to the "Absolute": Triad is the most fundamental structure in human society, *Visnik of the Academy of Science of Ukraine*, 1993, No. 5, pp. 29-35 (in Ukrainian).

Burgin, M. (1993b) Information triads, *Philosophical and Sociological Thought*, No. 7-8, pp. 243-246 (in Russian and Ukrainian).

Burgin, M. (1994) Fundamental Base of the Theory of Triads, *Idea*, No. 2, pp. 32-45 (in Ukrainian).

Burgin, M. (1994a) Is it Possible that Mathematics Gives new Knowledge about Reality? *Philosophical and Sociological Thought*, No. 1, pp. 240-249 (in Russian and Ukrainian).

Burgin, M. (1994b) Structural traits of psychoanalysis, *Visnik of the National Academy of Sciences of Ukraine*, No. 2, pp. 45-50 (in Ukrainian).

Burgin, M. (1995) Named Sets as a Basic Tool in Epistemology, *Epistemologia*, v. XVIII, pp. 87-110.

Burgin, M. (1995a) The Phenomenon of Knowledge, *Philosophical and sociological Thought*, No. 3-4, pp. 41-63 (in Russian and Ukrainian).

Burgin, M. (1996) *The Structural Level of Nature*, Znannya, Kiev (in Russian).

Burgin, M. (1996a) Triad as a Fundamental Psychological Structure, in *Mathematics and Psychology in the Pedagogical System "Technical University"*, Odessa, 1996, pp. 6-7 (in Russian).

Burgin, M. (1996b) Flow-charts in programming: arguments pro et contra, *Control Systems and Machines*, No. 4-5, pp. 19-29 (in Russian).

Burgin, M. *Fundamental Structures of Knowledge and Information*, Ukrainian Academy of Information Sciences, Kiev, 1997 (in Russian).

Burgin, M. (1997a) The Triad of Power, in *Politological Encyclopedic Dictionary*, Kyiv, pp. 353-354 (in Ukrainian).

Burgin, M. (1997b) Scientific Foundations of Structuralism, in *Language and Culture*, v.1, Kiev, pp. 24-25 (in Russian).

Burgin, M. (1997/98) Empirical Foundations of the Theory of Triads, *Reports of the International Convent of the Trinitary Knowledge*, No. 1, pp. 119-127 (in Ukrainian).

Burgin, M. *On the Nature and Essence of Mathematics*, Ukrainian Academy of Information Sciences, Kiev, 1998 (in Russian).

Burgin, M. (1998a) The Structure of Culture in Modern Society, in *Culture as a Living Whole*, Odessa, pp. 26-28 (in Russian).

Burgin, M. *Intellectual Components of Creativity*, AACS, Kiyv, 1998b (in Ukrainian)

Burgin, M. *Unified Foundations of Mathematics*, Preprint Mathematics LO/0403186, 2004, 39 p. (electronic edition: http://arXiv.org)

Burgin, M. (2004a) Data, Information, and Knowledge, *Information*, v. 7, No.1, pp. 47-57.

Burgin, M. *Super-recursive Algorithms*, Springer, New York/Heidelberg/Berlin, 2005.

Burgin, M. (2006) Mathematical Schema Theory for Modeling in Business and Industry, Proceedings of the 2006 *Spring Simulation MultiConference* (SpringSim '06), Huntsville, Alabama, pp. 229-234.

Burgin, M. (2008) Structural Organization of Temporal Databases, in *Proceedings of the 17th International Conference on Software Engineering and Data Engineering* (SEDE-2008), ISCA, Los Angeles, California, pp. 68-73.

Burgin, M. (2009) Structures in mathematics and beyond, in Proceedings of the 8th *Annual International Conference on Statistics, Mathematics and Related Fields*, Honolulu, Hawaii, pp. 449-469.

Burgin, M. *Theory of Information: Fundamentality, Diversity and Unification*, World Scientific, New York/London/Singapore, 2010.

Burgin, M. (2010a) Mathematical Schema Theory for Network Design, in Proceedings of the ISCA 25th International Conference "*Computers and their Applications*" (CATA-2010), ISCA, Honolulu, Hawaii, pp. 157-162.

Burgin, M. (2010b) Structure Integration, *Integration: Mathematical Theory and Applications*, v. 2, No. 2, pp. 167-212.

Burgin, M. *Theory of Named Sets*, Nova Science Publishers, New York, 2011.

Burgin, M. (2011a) Information in the Structure of the World, *Information: Theories & Applications*, 2011, v.18, No. 1, pp. 16 - 32.

Burgin, M. and Burgina, E. (1982) Partitions in Languages and Parallel Computations, *Programming and Computer Software*, v. 8, No. 3, pp. 112-120 (translated from Russian).

Burgin, M. and Burgina, E. (1982a) Information retrieval and multi-valued partitions in languages, *Cybernetics and System Analysis*, No. 1, pp. 30-42 (translated from Russian).

Burgin, M. and Chelovsky, Y. (1982) Multidimensional structural analysis of FORTRAN programs, *Control Systems and Machines*, No. 1, pp. 38-39 (in Russian).

Burgin, M. and Chelovsky, Y. (1984) Program Quality Estimation Based on Structural Characteristics, in *Proceedings of the International Conference "Software"*, Kalinin, pp. 79-81 (in Russian).

Burgin, M. and Eggert, P. (2004) Types of Software Systems and Structural Features of Programming and Simulation Languages, in *Proceedings of the Business and Industry Simulation Symposium*, Society for Modeling and Simulation International, Arlington, Virginia, pp. 177-181.

Burgin, M. and Gladun, V.P. (1989) Mathematical Foundations of the Semantic Networks Theory, *Lecture Notes in Computer Science*, v. 364, pp. 117-135.

Burgin, M. and Gorsky, D. (1991) Towards the construction of general theory of concept, in *The Opened Curtain*, Oulder/San Francisco /Oxford, pp. 167-195.

Burgin, M. and Karasik, A. (1976) Operators of Multidimensional Structured Model of Parallel Computations, *Automation and Remote Control*, v. 37, No. 8, pp. 1295-1300 (translated from Russian).

Burgin, M. and Kuznetsov, V. (1991) Structure and Development of Mathematical Theories, *Modern Logic*, v.2, No. 1, pp. 3-28

Burgin, M. and Kuznetsov, V. *The Structure-Nominative Analysis of Theoretical Knowledge*, Proceedings of the Institute of Philosophy, Kiev, 1992

Burgin, M. and Kuznetsov, V. (1992a) The Structure-Nominative Direction in Methodology of Science (1984-1991), in *Methodological Conceptions and Schools in the USSR* (1951-1991), Novosibirsk, pp. 111-130 (in Russian)

Burgin, M. and Kuznetsov, V. *Nomological structures in Scientific Theories*, Naukova Dumka, Kiev, 1993 (in Russian)

Burgin, M. and Kuznetsov, V. *Introduction to Modern Exact Methodology of Science*, International Science Foundation, Moscow, 1994 (in Russian) .

Burgin, M. and Milov, Yu. (1998) Grammatical Aspects of Language in the Context of the Existential Triad Concept, in *On the Nature and Essence of Mathematics*, Ukrainian Academy of Information Sciences, Kiev, pp. 136-142.

Burgin, M.S. and Neishtadt, L.A. (1992) *Structured psychological models of man-computer communication*, International Conference on Bionic Modeling, S.-Petersburg, pp. 157-158.

Burgin, M. and Neishtadt, L. *Communication and discourse in teachers professional activity*, Daugavpils Pedagogical Institute, Daugavpils, 1993

Burgin, M. and Schumann, J. (2006) Three Levels of the Symbolosphere, *Semiotica*, v. 160, No. 1/4, pp. 185-202.

Burgin, M. and Zak, Y. (1982) Principles of automated system design, *Control Systems and Machines*, No. 2, pp. 82-89 (in Russian).

Burgin, M. and Zak, Y. (1984) Structure-parametric description of graphic objects, *Control Systems and Machines*, No. 3, pp. 80-85 (in Russian).

Burton, D.M. *The History of Mathematics*, The McGrow Hill Co., New York, 1997

Button, T. (2006) Realist Structuralism's Identity Crisis: A Hybrid Solution, *Analysis*, v. 66, pp. 216–22.

Caianiello, E.R., Marinaro, M., Scarpetta G. and Simocelli, G. (1988) Structure and Modularity in Self-organizing Complex Systems, in *Research in Structure Theory*, Nauka, Moscow, pp. 144-176 (in Russian).

Capuro, R., and Hjorland, B. (2003) The Concept of Information, *Annual Review of Information Science and Technology*, v. 37, No. 8, pp. 343-411

Caraway, C. (2003) Wittgenstein on the Structure of Justification: Breaking New Epistemological Ground, *Proc. Austrian Ludwig Wittgenstein Society*, pp. 78-80.

Carlip, S. Mosna, R. A. and Pitelli, J. P. M. (2011) Vacuum Fluctuations and the Small Scale Structure of Spacetime, *Phys. Rev. Lett.*, v. 107, 021303.

Carnap, R. *Der logische Aufbau der Welt. Scheinprobleme in der Philosophie*, Berlin, 1928 (English translation: *The Logical Structure of the World. Pseudoproblems in Philosophy*, London/Berkeley, CA, 1967).

Carter, J. (2008) Structuralism as a Philosophy of Mathematical Practice, *Synthese*, v. 163, No. 2, pp. 119-131.

Casati, R. and Varzi, A. *Parts and Places: the structures of spatial representation*, MIT Press, Cambridge, MA, 1999.

Castell, L., Drieschner, M., and von Weizsäcker, C.F. (Eds.) (1975-1986) *Quantum Theory and the Structures of Time and Space* (6 volumes), Hanser, Munich.

Caws, P. *Structuralism: The Art of the Intelligible*, Humanities Press, Atlantic Highlands, NJ, 1988.

Chakravartty A. (1998) Semirealism, *Studies in History and Philosophy of Modern Science*, v. 29, pp. 391–408.

Chakravartty, A. (2003) The Structuralist Conception of Objects, *Philosophy of Science*, v. 70, pp. 867–878.

Chakravartty, A. (2004) Structuralism as a Form of Scientific Realism, *International Studies in Philosophy of Science*, v. 18, pp. 151–171.

Chakravartty, A. *A Metaphysics for Scientific Realism: Knowing the Unobservable*. Cambridge University Press, Cambridge, 2007.

Church, A. A (1951) Formulation of the Logic of Sense and Denotation, in *Structure, Method and Meaning: Essays in Honor of Henry M. Sheffer*, Liberal Arts Press, New York, pp. 3-24.

Church, A. *Introduction to Mathematical Logic*, Princeton University Press, Princeton, 1956.

Clement, W.C. (1953) Russell's Structuralist Thesis, *The Philosophical Review*, v. 62, No. 2, pp. 266-275

Codd, E. *Relational Model for Data Management*, Addison-Wesley, Reading, MA, 1990.

Cohen, B. and Murphy, G.L. (1984) Models of concepts, *Cognitive Science*, v. 8, No. 1, pp. 27-58.

Cohn, P.M. *Universal algebra*, Harper & Row, New York/ Evanston/London, 1965.

Cook, K. (2000) Charting Futures for Sociology: Structure and Action, *Contemporary Sociology*, v. 29, pp. 685-692.

Corry, L. *Modern Algebra and the Rise Mathematical Structures*, Birkhäuser, Basel/Boston/Berlin, 1996

Cottrell, A.H. and Pettifor, D.G. (2000) Models of Structure, in *Structure in Science and Art*, Cambridge University Press, Cambridge, pp. 33-47.

Cumming, S. (2009) Names, in *Stanford Encyclopedia of Philosophy* http://plato.stanford.edu/.

Daft, R.L. *Organization Theory and Design*, International Thomson P.C., Cincinnati, Ohio, 1997.

Dahl, O.J., Dijkstra, E.W. and Hoare, C.A.R. *Structured Programming*, Academic Press, London/New York, 1972.

Date, C. J. *An Introduction to Database Systems*, Addison-Wesley, Boston/San Francisco/New York, 2004.

Dedekind, R. *Was sind und was sollen die Zahlen?* Vieweg und Sohn, Braunschweig, 1888.

Devlin, K. *Mathematics: The Science of Patterns*, Scientific American Library, New York 1994.

Ding, E. (2009) The Platonic Triad and Its Chinese Counterpart, *Signs*, v. 3, pp. 41-56.

Dirac, P. A. M. (1928) The Quantum Theory of the Electron, *Proceedings of the Royal Society of London*, Series A, v. 117, No. 778, pp. 610–624

Dirac, P.A.M. *A Theory of Electrons and Protons*, Proceedings of the Royal Society of London, A 126 (5), 1930, pp. 360-365

Dirac, P. A. M. (1931) Quantised Singularities in the Quantum Field, *Proc. R. Soc. Lond. A*, v. 133, No. 821, pp. 2–3

Dirac, P.A.M. (1983) The origin of quantum field theory, in *The Birth of Particle Physics*, Cambridge University Press, Cambridge, pp. 39-55.

Dorato, M. (2000) Substantivalism, Relationism and Structural Spacetime Realism, *Foundations of Physics*, vol 30 (10): 1605-1628

Dosse, F. *History of Structuralism*, University of Minnesota Press, Minneapolis, Minnesota, 1998.

Duhem, P. *La théorie physique, son objet, sa structure*, Paris 1906 (English translation: *The Aim and Structure of Physical Theory*, Princeton University Press, Princeton, 1954).

Dumas, J. B. *Affinite* (Lecture Notes), vii, Academie des Sciences, Paris, 1837.

Durkheim, E. *The Division of Labor in Society*, The Free Press, New York, 1984.

Dutt, A. K. and Ross, J. (Eds.) *Development Economics and Structuralist Macroeconomics*, Essays in honor of Lance Taylor, Edward Elgar, 2003

Eco, U. *A Theory of Semiotics*, Indiana University Press, Bloomington, 1976.

Eco, U. *Semiotics and the philosophy of language*, Indiana University Press Bloomington, 1984.

Eddington, A.S. *The Theory of Relativity and its Influence on Scientific Thought*, Clarendon Press, Oxford, 1922

Elmasri, R. and Navathe, S.B. *Fundamentals Database Systems*, Addison-Wesley Publishing Company, Reading, Massachusetts, 2000.

Erlich, V. *Russian Formalism*, Yale University Press, Yale, 1981.

Eschenbach, C. (1995) Classical mereology and restricted domains, *International journal of human-computer studies*, v. 43, pp. 723-740.

Esfeld, M. and Lam, V. (2006) Moderate Structural Realism About Space-Time, *Synthese*, v. 160, No. 1, pp. 27 - 46

Feferman, S. (1977) Categorical Foundations and Foundations of Category Theory, in *Logic, Foundations of Mathematics and Computability Theory*, Dordrecht: Reidel, pp. 149–69.

Ferré, S. and Rudolph, S. (Eds.) *Formal Concept Analysis*, Proceedings of the 7th International Conference, ICFCA 2009, Darmstadt, Germany, 2009, LNCS 5548, 2009.

Fichte, J.G. *Grundlage der gesamten Wissenschaftslehre*, Ch. R. Gabler, 1794 (English translation: Fichte, J.G. *Introduction to the* Wissenschaftslehre *and Other Writings (1797 - 1800)*, Hackett, Indianapolis, IN, 1994).

Fisher, R.A. (1922) On the mathematical foundations of theoretical statistics. *Philos. Trans. Roy. Soc., London, Sec. A*, v. 222, pp. 309-368.

Fiske, A. P. *Structures of social life*: *The four elementary forms of human relations*, Free Press, New York, 1991.

FitzGerald, E. V. K. (1990) Kalecki on Financing Development: An Approach to the Macroeconomics of the Semi-industrialised Economy, *Cambridge Journal of Economics*, v. 14, No. 2, pp. 183-203.

Flükiger, D.F. (1999) Towards a unified concept of information: Presentation of a new approach, in *The quest for a unified theory of information. Proceedings of the Second International Conference on the Foundations of Information Science*, Gordon and Breach, Amsterdam, pp. 101-111

Foradori, E. *Grundgedanken der Teiltheorie*, Leipzig, Germany, 1937.

Freedman, D. and Humphreys, P. (1999) Are there Algorithms that Discover Causal Structure? *Synthese*, v. 121, pp. 29-54.

Frege, G. *Funktion und Begriff*, Hermann Pohle, Jena, 1891.

Frege, G. (1892) Über Begriff und Gegenstand, *Vierteljahrsschrift für wissenschaftliche Philosophie*, v. 16, pp. 192-205.

Frege, G. (1892a) Über Sinn und Bedeutung, *Zeitschrift für Philosophie und philosophische Kritik*, v. 100, pp. 25-50.

French, S. (2000) The Reasonable Effectiveness of Mathematics: Partial structures and the application of group theory to physics, *Synthese*, v. 125, No. 1-2, pp. 103-120

French, S. (2011) Shifting to Structures in Physics and Biology: A Prophylactic for Promiscuous Realism, *Studies in History and Philosophy of Science*, Part C, v. 42, No. 2, pp. 164-173.

Freud, S. *The Ego and the Id*, The Hogarth Press Ltd, London, 1949.

Frieden, R.B. *Physics from Fisher Information*, Cambridge University Press, Cambridge, 1998.

Friedman, J.P. *Dictionary of Business Terms*, Barron's Educational Series, 2000.

Frigg, R. and Votsis, I. (2011) Everything you always wanted to know about structural realism but were afraid to ask, *Euro Journal of Philosophy of Science*, v. 1, pp. 227–276.

Gaifman, H. (2006) Naming and Diagonalization, from Cantor to Gödel to Kleene, *Logic Journal of IGPL*, v. 14, No. 5, pp. 709-728.

Gandy, R. (1973) "Structure" in Mathematics' in *Structuralism: An introduction*, Oxford: Clarendon Press, pp 138-153

Germain, P. (1971) A general view of the evolution of mathematics, in *Great Currents of Mathematical Thought*, Dover Publications, New York, pp. 231-248.

Giddens, A. *The Constitution of Society*: *Outline of the Theory of Structuration*, Polity Press, Cambridge, 1984.

Glansdorff, P. and Prigogine, I. *Thermodynamic Theory of Structure, Stability and Fluctuations*, Wiley-Interscience, London, 1971.

Goldberger, A. (1972). Structural equation models in the social sciences, *Econometrica*: *Journal of the Econometric Society*, v. 40, pp. 979–100.

Goldblatt, R. *Topoi: The Categorical analysis of Logic*, North-Holland P.C., Amsterdam, 1984.

Goodman, N., *The Structure of Appearance*, Cambridge, Massachusetts: Harvard University Press, 1951

Gopnik, A. and Meltzoff, A.N. (1992) Categorization and Naming: Basic-Level Sorting in Eighteen-Month-Olds and its Relation to Language, *Child Development*, v.63, No. 5, pp. 1091-1103.

Goppold, A. (1998) Information and third order ontology, *Biosystems*, v. 46, No. 1-2, pp. 169-173.

Gower, B. (2000) Cassirer, Schlick and 'structural' realism: The philosophy of the exact sciences in the background to early logical empiricism, *British Journal for the History of Philosophy*, v. 8, pp. 71–106.

Grossmann, R. (1990) *The Fourth Way: A Theory of Knowledge*, Indiana University Press, Bloomington/Inianapolis.

Gryakalov, A.A. *Structuralism in Esthetics*, Leningrad University Press, Leningrad, 1989 (in Russian).

Gutting, G. (1973) Conceptual Structures and Scientific Change, *Studies in the History and Philosophy of Science*, v. 4/3, pp. 209-230.

Halliday, M.A.K. (1967) Notes on Transitivity and Theme in English, *Journal of Linguistics*, v. 3, pp. 199-244

Hampton, J.A. (2011) Concepts and Natural Language, in *Concepts and Fuzzy Logic*, MIT Press, Cambridge, MA, pp. 233-258.

Hartmann, N. *Der Aufbau der realen Welt. Grundriss der allgemeinen Kategorienlehre*, Walter De Gruyter, Berlin, 1940.

Hartmann, N. *New Ways of Ontology*, Greenwood Press, Westport, 1952.

Heim, M. *The Metaphysics of Virtual Reality*, Oxford University Press, New York, 1993.

Heisenberg, W. *Philosophie - Le manuscrit de* 1942, Seuil, Paris, 1998.

Hellman, G. (1996) Structuralism Without Structures, *Philosophia Mathematica*, v. 4, No. 3, Special issue on Mathematical Structuralism, pp. 148-173.

Hellman, G. (2001) Three Varieties of Mathematical Structuralism, *Philosophia Mathematica*, v. 9, pp. 184–211.

Hellman, G. (2003) Does Category Theory Provide a Framework for Mathematical Structuralism? *Philosophia Mathematica*, v. 11, pp. 129–57..

Hellman, G. (2005) Structuralism, in The Oxford Handbook of Philosophy of Mathematics and Logic, Oxford University Press, New York, pp. 536–62..

Hellman, G. (2006) What is Categorical Structuralism? in *The Age of Alternative Logics: Assessing Philosophy of Mathematics and Logic Today*, Berlin: Springer-Verlag, pp. 131–62.

Hellman, G. *Foundational Frameworks, in Foundational Theories of Classical and Constructive Mathematics*, Springer-Verlag, Berlin, 2010.

Hennessy, J.L. and Patterson, D.A. *Computer Architecture: A Quantitative Approach*, Morgan Kaufmann, San Francisco/New York, 2007.

Henry, D. *Medieval Mereology*, Grüner, Amsterdam, 1991.

Henzinger, T.A. (2007) Quantitative Generalizations of Languages, *Lecture Notes in Computer Science*, v. 4588, pp. 20-22.

Herbert, N. *Quantum Reality: Beyond the New Physics*, Anchor Books, New York, 1987.

Hesse, M. B. *The Structure of Scientific Inference*, Macmillan, London, 1974.

Heyman, J. *Structural Analysis: A Historical Approach*, Cambridge University Press, Cambridge, 1998.

Heyman, J. *The Science of Structural Engineering*, Imperial College Press, London, 1999.

Hjelmslev L. (1958) Dans quelle mesureles significations des mots peuvent'elles etre consideres corn me formant une structure? In *Proceedings of the 8th International Congress of Linguistics,* Oslo, pp. 636-654.

Hjelmslev, L. *Prolegomena to a Theory of Language*, University of Wisconsin Press, Madison, 1963.

Hofmann, A., Wang, C.K., Osman, A. and Camp, D. (2010) Merging Structural Biology with Chemical Biology: Structural Chemistry at Eskitis, *Struct. Chem.* , v. 21, pp. 1117-1129

Holmes, M. R. (2001) Foundations of mathematics in polymorphic type theory, *Topoi*, v. 20, pp. 29-52.

Honda, K. (2000) Elementary structures in process theory: Sets with renaming, *Mathematical Structures in Computer Science*, v. 10, No. 5, pp. 617-663.

Hoyt, H. *The structure and growth of residential neighborhoods in American cities*, Federal Housing Administration, Washington DC, 1939.

Hsiao, K.-y., Tellex, S., Vosoughi, S., Kubat, R. and Roy, D. (2008) Object schemas for grounding language in a responsive robot, *Connect. Sci.*, v. 20, No. 4, pp. 253-276.

Hsiao, K.-y., Vosoughi, S., Tellex, S., Kubat, R. and Roy, D. (2008a) Object schemas for responsive robotic language use, in *HRI 2008*, pp. 233-240.

Husemöller, D. *Fibre Bundles*, Springer Verlag, Berlin/ Heidelberg /New York, 1994.

Jeyasingh, S. (2000) The Body in Space: Structures in Dance, in *Structure in Science and Art*, Cambridge University Press, Cambridge, pp. 48-72.

Ji, S. (2001) Isomorphism between Cell and Human Languages: Micro- and Macrosemiotics, in *Semiotics* 2000: "*Sebeok's Century*", Legas, New York. pp. 357-373.

Ji, S. (2002) Microsemiotics of DNA. *Semiotica*, v. 138, No. 1/4, pp. 15-42.

Ji, S. *Molecular Theory of the Living Cell: Concepts, Molecular Mechanisms, and Biomedical Applications*, Springer, New York, 2012.

Jones, R.B. *Membership Structures*, 2007 (electronic publication: http://www.rbjones.com/rbjpub/isar/HOL/Membership.pdf).

Jung, C.G. (1969) *The Structure and Dynamics of the Psyche*, Princeton University Press, Princeton.

Ketchum, R. J. (1979) Names, Forms and Conventionalism: *Cratylus*, pp. 383-395, *Phronesis*, v. 24, pp. 133-47.

King, J. C. (1995) Structured Propositions and Complex Predicates, *Nous*, v. 19, pp. 516-535.

King, J. C. (2011) Structured Propositions, *The Stanford Encyclopedia of Philosophy* http://plato.stanford.edu/.

Kleene, S.C. *Mathematical Logic*, John Wiley & Sons, Inc., New York, NY, 1967.

Kleiner, I. (2001) Modern Algebra and the Rise of Mathematical Structures, *Historia Mathematica*, v. 28, pp. 304-312.

Knuth, D. *The Art of Computer Programming*, v.1: *Fundamental Algorithms*, Addison-Wesley, Boston/San Francisco/New York, 1997.

Kolmogorov, A.N. (1953) On the Concept of Algorithm, *Russian Mathematical Surveys*, v. 8, No. 4, pp. 175-176 (in Russian).

Koslicki, K. *The Structure of Objects*, Oxford University Press, New York, 2008

Krause, D. (2005) Structures and structural realism, *Logic Journal of the IGPL*, v. 13, No. 1, pp. 113-126.

Kripke, S. *Naming and Necessity*, Basil Blackwell, Boston, 1980.

Kripke, S. (1992) Individual Concepts: Their Logic, Philosophy and Some of Their Uses, *Proceedings and Addresses of the APA*, v. 66, pp. 70-73.

Kuhn, T. S. *The Structure of Scientific Revolutions*, University of Chicago Press, Chicago, 1962.

Kuhn, T. S. (1981) Theoriewandel als Strukturwandel, in *Wissenschaftsgeschichte und wissenschaftliche Revolutionen*, Köln, pp. 114 -135..

Kulagina, O.S. (1958) One method of defining grammatical concepts on the basis of set theory, *Probl. Kibern.*, No. 1, pp. 203-214 (in Russian).

Küppers, B.-O. (2000) Die Strukturwissenschaften als Bindeglied zwischen Natur- und Geisteswissenschaften, in *Die Einheit der Wirklichkeit*, München, pp. 89-105.

Kuratowski, K. *Topology*, Academic Press, Waszawa, v. 1, 1966; v. 2, 1968.

Kurosh, A.G. *Lectures on general algebra*, Chelsea P. C., New York, 1963.

Lacan, J. (1977) *The Four Fundamental Concepts of Psychoanalysis*, Hogarth, London.

Ladyman, J. (1998) What is Structural Realism? *Studies in History and Philosophy of Science*, v. 29, pp. 409-424

Ladyman, J. *Structural Realism*, Stanford Encyclopedia of Philosophy, 2009 http://plato.stanford.edu/entries/structural-realism/.

Laughlin, C. D. and d'Aquili, E. G. *Biogenetic Structuralism*, Columbia University Press, New York, 1974.

Lambek, J. (1958) The mathematics of sentence structure, *American Mathematical Monthly*, v. 65, No. 3, pp. 154-170.

Landry, E. (1999) Category Theory as a Framework for Mathematical Structuralism, *The 1998 Annual Proceedings of the Canadian Society for the History and Philosophy of Mathematics*, pp. 133-142.

Landry, E. (2006) Category Theory as a Framework for an *in re* Interpretation of Mathematical Structuralism, in *The Age of Alternative Logics: Assessing Philosophy of Logic and Mathematics Today*, Springer Netherlands, pp. 163-179.

Landry, E.M. (2007) Shared Structure need not be Set-Structure, *Synthese*, v. 58, pp. 1–17.

Lao-Tzu's Taoteching, Translated by Porter, B. (aka Red Pine), Copper Canyon Press, Port Townsend, WA, 1996

Lautman, A. *Essai sur les notions de structure et d' existence en mathématique*, Hermann, Paris, 1938.

Lee, E. and Pakstas, A. (2002) Overview of computer-based virtual reality, in *Virtual reality technologies for future telecommunication systems*, pp. 3–7.

Leonard, H. S. and Goodman, N. (1940) The Calculus of Individuals and Its Uses, *Journal of Symbolic Logic*, v. 5 pp. 45-55.

Leśniewski, S. Podstawy ogólnej teoryi mnogosci, I, *Prace Polskiego Kola Naukowego w Moskwie*, Sekcya matematyczno-przyrodnicza, 1916 (Eng. trans. by D. I. Barnett: *Foundations of the General Theory of Manifolds* I, in S. Leśniewski, *Collected Works*, Dordrecht: Kluwer, v. 1, 1992, pp. 129-173).

Leśniewski, S. (1929) Grundzüge eines neuen Systems der Grundlagen der Mathematik, *Fundamenta Mathematicae*, v. 14, pp. 1–81.

Lévi-Strauss, C. *Anthropologie structurale*, Plon, Paris, 1958 (English translation: *Structural Anthropology*, Basic Books, New York, 1963).

Levi-Strauss, C.*The Elementary Structures of Kinship*, Beacon Press, 1971.

Lévi-Strauss, C. (1987) *Anthropology and Myth*: *Lectures, 1951-1982*, Blackwell, Oxford.

Levich, A. P. (1978) Information as a system structure, *Semiotika i Informatika*, v. 10, pp.116-132 (in Russian)..

Lewis, G.N. (1916) The atom and the molecule, *Journal Amer. Chem. Soc.*, v. 38, pp. 762-785.

Losev, A.F. *Notes on Antique Symbolism and Mythology*, Mysl, Moscow, 1993 (in Russian).

Lyre, H. *Quantentheorie der Information*, Springer, Vienna, 1998

MacKay, D. M. *Information, Mechanism and Meaning*, The MIT Press, Cambridge, Massachusetts, 1969.

MacLane, S. *Mathematics, Form and Function*, Springer-Verlag, New York/Boston/Berlin, 1986.

MacLane, S. (1996) Structure in mathematics, *Philosophia Mathematica*, v. 4, No. 3, pp. 174-183.

Magnavita, J. J. *Theories of Personality*: *Contemporary Approaches to the Science of Personality*, John Wiley & Sons, New York, 2002.

Mahoney, M.S. (2010) The Structures of Computation and the Mathematical Structure of Nature, *The Rutherford Journal*, v. 3 (electronic publication: http://www.rutherford journal.org/current2.html).

Mal'cev, A.I. *The metamathematics of algebraic systems*, North-Holland, Amsterdam, 1971.

Manin, Y.I. *Georg Cantor and his Heritage*, Preprint in mathematics, math.AG/0209244, 2002 (electronic edition: http://arXiv.org).

Margolis, E. (1994) A Reassessment of the Shift from Classical Theories of Concepts to Prototype Theory, *Cognition*, v. 51, pp. 73-89.

Margolis, E. and Laurence, S. (Eds.) *Concepts*: *Core Readings*, MIT Press, Cambridge, Massachusetts, 1999.

Martin, W. *Structures*: *Theory and Analysis*, Palgrave Macmillan, London, 1999.

Maxwell, G. (1968) Scientific methodology and the causal theory of perception, in *Problems in the philosophy of science*, North-Holland Publishing Company, Amsterdam.

Maxwell, G. (1970) Theories, Perception and Structural Realism, in *The Nature and Function of Scientific Theories* (University of Pittsburgh Series in the Philosophy of Science, v. 4), University of Pittsburgh Press, Pittsburgh, pp. 3–34.

McArthur, D. (2006) Recent Debates Over Structural Realism, *Journal for General Philosophy of Science* , v. 37, No. 2, pp.209 - 224.

McCulloch, G. *The Game of the Name*: *Introducing Logic, Language and Mind*, Oxford University Press, Oxford, 1989.

McDowell, J. (1977) On the Sense and Reference of a Proper Name, *Mind*, v. 86, pp. 159–185.

McGraw-Hill Encyclopedia of Science and Technology, 5th Edition, McGraw-Hill Companies, 2005.

McLendon, H. J. (1955) Uses of Similarity of Structure in Contemporary Philosophy, *Mind*, v. 64, pp. 79-95

McMullin, E. (1990) Comment: Duhem's middle way, *Synthese*, v. 83, pp. 421–430.

Medin, D.L., and Ortony, A. (1989) Psychological essentialism, in *Similarity and Analogical Reasoning*, Cambridge University Press, Cambridge, pp. 179-195.

Medin, D.L., and Shoben, E. J. (1988) Context and Structure in Conceptual Combination, *Cognitive Psychology*, v. 20, pp. 158-190.

Menger, K. (1940) Topology without points, *Rice Inst. Pamphlet*, v. 27, pp. 80-107.

Mesarovic, M. D., and Takahara, Y. *General Systems Theory: Mathematical Foundations*, Academic Press, New York/London/San Francisco, 1975.

Michalewicz, Z. *Genetic Algorithms + Data Structures = Evolution Programs*, Springer-Verlag, New York, 1996.

Milgram, P. and Kishino, A. F. (1994) A Taxonomy of Mixed Reality Visual Displays, *IEICE Transactions on Information and Systems*, E77-D(12), pp. 1321-1329.

Milgram, P., Takemura, H., Utsumi, A. and Kishino, F. (1994) Augmented Reality: A class of displays on the reality-virtuality continuum, in *Proceedings of Telemanipulator and Telepresence Technologies*. pp. 2351–2354.

Mill, J. S. *A System of Logic*, v. 1, Parker, Son, and Bourn, London, 1862.

Millar, A. (1991) Concepts, Experience, and Inference, *Mind*, C, No. 4, pp. 495-505..

Millar, A. (1994) Possessing Concepts, *Mind*, v. 103, pp. 73-81.

Mills, J. (2004) Faceted classification and logical division in information retrieval, *Information Trends*, v. 52, No. 3, pp. 542-570.

Mormann, T. (1985) Topologische Aspekte Strukturalistischer Rekonstruktionen, *Erkenntnis*, v. 23, No. 3, pp. 319 - 359.

Mormann, T. (1992) Structural Accessibility and Similarity of Possible Worlds, *Journal of Philosophical Logic*, v. 21, No. 2, pp.149 – 172.

Mormann, T. (1994) Accessibility, Kinds, and Laws: A Structural Explication, *Philosophy of Science*, v. 61, No. 3, pp. 389-406.

Morris, C. W. (1938) Foundation of the Theory of Signs, in *International Encyclopedia of Unified Science*, v. 1, No. 2.

Morris, C. W. *Signs, Language and Behavior*, Prentice-Hall, Inc., New York, 1946.

Morris, C. W. *Signification and Significance: A Study of the Relations of Signs and Values*, MIT Press, Cambridge, Massachusetts, 1964.

Mrozek, D. Malysiak-Mrozek, B. and Kozielski, S. (2009) Alignment of protein structure energy patterns represented as sequences of fuzzy numbers, in *Annual Meeting of the North American Fuzzy Information Processing Society* (NAFIPS 2009), Cincinnati, OH, pp. 1 - 6.

Mueller, B. and Rafelski, J. *The Structured Vacuum - Thinking about Nothing*, Verlag Harri Deutsch, 1985.

Muller, F.A. *Structures for Everyone*, A. Gerits & Son, Amsterdam, 1998.

Müller, U. (1994) Philosophie und Struktur: Über die Notwendigkeit strukturtheoretischer Reflexionen in den hermeneutischen Wissenschaften, *Zeitschrift für allgemeinemWissenschaftstheorie*, v. 25, pp. 261-277.

Murphy, G.L. and Medin, D. L. (1985) The role of theories in conceptual coherence, *Psychological Review*, v. 92, pp. 289-316.

Nguen, N. T. (2008) Inconsistency of knowledge and collective intelligence, *Cybernetics and Systems*, v. 39, No. 6, pp. 542-562.

Nicolescu, B. *Manifesto of Transdisciplinarity*, State University of New York Press, Albany, N.Y., 2002.

North, D. *Structure and Change in Economic History*, W.W. Norton, New York, 1981.

North, J. (2009) The "Structure" of Physics: A Case Study, *Journal of Philosophy*, v. 106, pp. 57-88.

Nöth, W. *Handbook of Semiotics*, Indiana University Press, Bloomington, IN, 1990.

Ogden C.K., and Richards I.A. *The Meaning of Meaning*, Routledge and Kegan, London, 1953

Ore, O. (1935) On the foundation of abstract algebra, I, *Ann. of Math.*, v. 36, pp. 406-437.

Ore, O. (1936) On the foundation of abstract algebra, II, *Ann. of Math.*, v. 37, pp. 265-292.

Osherson, D.N., and E. E. Smith. (1981) On the adequacy of prototype theory as a theory of concepts, *Cognition*, v. 11, No. 35-58

Palter, R. (1956) Physics and Structure, *Philosophical Review*, v. 65, No. 3, pp. 371-384

Parsons, C. (1990) The Structuralist View of Mathematical Objects, *Synthese*, v. 84, pp. 303-346.

Parsons, T. *The Structure of Social Action*, The Free Press, New York, 1968.

Paul, P., Fleig, O. and Jannin, P. (2005) Augmented virtuality based on stereoscopic reconstruction in multimodal image-guided neurosurgery: methods and performance evaluation, *IEEE Trans Med Imaging*, v. 24, No. 11, pp. 1500-1511.

Peirce, C. S. *A Syllabus of Certain Topics of Logic*, Alfred Mudge & Son, Boston, 1903.

Peirce C. S. *Collected papers*, v. 1-6, Cambridge University Press, Cambridge, England, 1931-1935.

Penrose, R. *Shadows of the Mind*, Oxford University Press, Oxford, 1994.

Petrovsky, A.B. (1997) Structuring Techniques in Multiset Spaces, in *Multiple Criteria Decision Making*, Lecture Notes in Economics and Mathematical Systems, v. 448, Springer-Verlag, Berlin, pp. 174-184.

Piaget, J. (1964) Mother structures and the notion of number, in *Cognitive Studies and Curriculum Development*, School of Education, Cornell University, pp. 33-39.

Piaget, J. *Structuralism*, Routledge and Kegan Paul, London, 1971.

Pitts, A. M. (2011) Structural Recursion with Locally Scoped Names, *Journal of Functional Programming*, v. 21, No. 3, pp. 235-286.

Pleiffer, D. and Niehaves, B. (2005) Evaluation of conceptual models – a structuralist approach, *Proceedings of the European Conference on Information Systems* (ECIS 2005) (electronic publication: http://aisel.aisnet.org/ecis2005/43/).

Plato, *The Collected Dialogues of Plato*, Princeton University Press, Princeton, 1961.

Poincaré, H. *La Science et l'hypothèse*, Flammarion, Paris, 1902.

Poincaré, H. *La valeur de la science*, Flammarion, Paris, 1905.

Poincaré, A. *Scince et Méthode*, Flamarion, Paris, 1908.

Poli, R. (2001) The Basic Problem of Levels of Reality, *Axiomathes*, v. 12, pp. 3-4..

Poli, R. (2003) Descriptive, Formal and Formalized Ontologies, in *Husserl's Logical Investigations Reconsidered*, Kluwer Academic, Dordrecht.

Poli, R. (2006) The Theory of Levels of Reality and the Difference between Simple and Tangled Hierarchies, *Systemics of Emergence: Research and Development*, v. 10, pp. 715-722.

Popper, K. R. (1974) Replies to my critics, in *The Philosophy of Karl Popper*, Open Court, La Salle, IL, pp. 949-1180.

Popper, K. *Unended Quest: An Intellectual Autobiography*, Open Court Pub Co., Chicago, IL, 1982.

Popper, K. R. *Objective knowledge: An evolutionary approach*, Oxford University Press, New York, 1979.

Porpora, D. V. *The Concept of Social Structure*, Greenwood Press, New York/ Wetport/London, 1987

Propp, V. *Morphology of the Folktale*, Leningrad, 1928 (English translation: Mouton, The Hague, 1958).

Pullan, W. and Bhadeshia, H. (Eds.) *Structure in Science and Art*, Cambridge University Press, Cambridge, 2000.

Putnam, H. (1975) The Meaning of 'Meaning', in *Language, Mind, and Knowledge, Minnesota Studies in the Philosophy of Science* (Volume 7), University of Minnesota Press, Minneapolis, Minnesota, pp. 131–193.

Quine, W. V. O. *Word and Object: An inquiry into the linguistic mechanisms of objective reference*, The M.I.T. Press, Cambridge, MA, 1960.

Quine, W.V. (1992) Structure and Nature, *Journal of Philosophy*, v. 89, pp. 5-9..

Rabin, M.O. (1969) Decidability of Second-order Theories and Automata on Infinite Trees, *Transactions of the AMS*, v. 141, pp. 1-35.

Reck, E. H. and Price, M. P. (2000) Structures and Structuralism in Contemporary Philosophy of Mathematics, *Synthese* 125, pp. 341-416.

Resnick, M. (1981) Mathematics as a science of patterns: Ontology and reference, *Nous*, v. 15, pp. 529-550.

Resnick, M. *Mathematics as a Science of Structures*, Oxford University Press, Oxford, 1997.

Resnick, M. D. *Mathematics as a Science of Patterns*, Clarendon Press, Oxford, 1999.

Resnikoff, H. L. *The illusion of reality*, Springer-Verlag, New York, 1989

Rickles, D., French, S. and Saatsi, J.T. (Eds.) *The Structural Foundations of Quantum Gravity*, Oxford University Press, Oxford, 2006

Ritzer, G. *Contemporary Sociological Theory and Its Classical Roots: The Basics*, Mcgraw-Hill, St. Louis, 2007.

Riva, G. (2009) Is presence a technology issue? Some insights from cognitive sciences, *Virtual Reality*, v. 13, pp.159–169.

Rivkin, S. and Sutherland, F. *The Making of a Name: The Inside Story of the Brands We Buy*, Oxford University Press, Oxford, 2004.

Roberts, C. (1996) Information Structure in Discourse: Towards an Integrated Formal Theory of Pragmatics, in *OSU Working Papers in Linguistics*, pp.91-136

Roberts, B. W. (2011) Group Structural Realism, *British Journal for the Philosophy of Science* , v. 62, No. 1, pp. 47-69.

Robertson, E.L. *An algebra of triadic relations*, Technical Report No. 606, Computer Science Department, Indiana University, Bloomigton, Indiana, 2005.

Robinson, A. *Introduction to Model Theory and Metamathematics of Algebra*, North-Holland, Amsterdam/New York, 1963.

Rocchi, P. *The Structural Theory of Probability: New Ideas from Computer Science on the Ancient Problem of Probability Interpretation*, Series in Computer Science, Plenum Publishers/Kluwer, New York, 2003.

Rombach, H. *Strukturontologie*, Alber, Freiburg/München, 1971.

Rombach, H. *Die Welt als Lebendige Struktur: Probleme und Lösungen Der Strukturontologie*, ach Verlag Kg, 2003.

Rosch, E. and Mervis, C. B. (1975) Family resemblances: studies in the internal structure of categories, *Cognitive Psychology*, v. 7, pp. 573-605

Rosner, J.A. (1996) Quine's Global Structuralism', *Dialectica*, v. 50, No. 3, pp. 235-242

Rucker, R. *Mind Tools: The Five Levels of Mathematical Reality*, Houghton Mifflin Co., Boston, 1987.

Ruse, M. (1973) *The Philosophy of Biology*, Hutchinson University Library, London.

Russell, B. *Principles of Mathematics*, Cambridge University Press, Cambridge, 1903.

Russell, B. (1905) On Denoting, *Mind*, v. 14, pp. 479-493.

Russell, B. (1908) Mathematical Logic as Based on the Theory of Types, *American Journal of Mathematics*, v. 30, pp. 222–262.

Russell, B. *The problems of philosophy*, Oxford University Press, Oxford, 1912.

Russell, B. *The Principles of Mathematical Philosophy*, George Allen and Unwin, London, 1919 .

Russell, B. *The Analysis of Matter*, George Allen and Unwin, London, 1927.

Russell, B. *Human Knowledge: Its Scope and Limits*, George Allen and Unwin, London, 1948.

Sainsbury, M. (1993) Russell on Names and Communication, in *Russell and Analytic Philosophy*, University of Toronto Press, Toronto, pp. 3–21.

Saltzer, J. (1978) Naming and Binding of objects, *Lecture Notes on Computer Science*, v. 60, pp. 99-208.

de Saussure, F. (1916) *Cours de linguistique générale*, ed. C. Bally and A. Sechehaye, with the collaboration of A. Riedlinger, Payot, Lausanne and Paris (English translation by W. Baskin: *Course in General Linguistics*, Fontana/Collins, Glasgow, 1977).

de Saussure, F. (1916a) Nature of the Linguistics Sign, in *Cours de linguistique générale*, McGraw Hill Education.

Scharffe, F. and Ding, Y. (2006) Three Levels of Knowledge Structuration for the Web, *Citeulike*, (electronic publication: http://www.citeulike.org/user/Sulpicus/article/1092431).

Scheffler, I. *The Anatomy of Inquiry*, Knopf, New York, 1963.

Schrödinger, E., *Science and humanism*, Cambridge University Press, New York, 1952

Searl, J. *Making the Social World: The Strture of Human Civilization*, Oxford University Press, Oxford, 2010.

Shapiro, S. (1989) Structure and Ontology, *Philosophical Topics*, v. 17, pp. 145–71.

Shapiro, S. *Philosophy of Mathematics: Structure and Ontology*, Oxford University Press, New York, 1997.

Shapiro, S. *Thinking about Mathematics: The Philosophy of Mathematics*, Oxford University Press, New York, 2000.

Shapiro, S. (2005) Categories, Structures, and the Frege-Hilbert Controversy: The Status of Metamathematics, *Philosophia Mathematica*, v. 13, pp. 61–73..

Shapiro, S. (2006) Structure and Identity, in *Identity and Modality*, Oxford University Press, New York, pp. 109–45.

Shapiro, S. (2008) Identity, Indiscernibility, and ante rem Structuralism: The Tale of *i* and –*i*, *Philosophia Mathematica*, v. 16, pp. 285–309.

Shapiro, S. Mathematical Structuralism, *Internet Encyclopedia of Philosophy*, 2010 (http://www.iep.utm.edu/m-struct/).

Shaumjan, S. K. *Principles of structural linguistics*, Mouton, The Hague, 1971.

Sherman, W.R., Craig, A.B. *Understanding virtual reality: interface, application, and design*, Elsevier, Amsterdam, 2003.

Shields, C. A Fundamental Problem about Hylomorphism, *Stanford Encyclopedia of Philosophy*, Stanford University, 2009 (http://plato.stanford.edu/).

Shoemaker, S. (1980) Causality and Properties, in *Time and Cause*, D. Reidel, Dordrecht.

Shoenflies, A. (1921) Zur Axiomatik den Mengenlehre, *Math. Annalen*, v. 83, pp. 173-200.

Simons, P. *Parts: A Study in Ontology*, Oxford University Press, Oxford, 1987.

Sinden, R. *DNA structure and function*, Elsevier, Amsterdam, 1994.

Skagestad, P. (1993) Thinking with machines: Intelligence augmentation, evolutionary epistemology, and semiotics, *Journal of Social and Evolutionary Systems*, v. 16, pp. 157-180.

Slupecki, J. Leśniewski's Podstawy ogólnej teoryi mnogosci, I, *Prace Polskiego Kola Naukowego w Moskwie*, Sekcya matematyczno-przyrodnicza, 1916.

Slupecki, J. (1955) Lesniewsjki's Calculus of Names, *Studia Logica* , v. 3, pp. 7-70.

Slupecki, J. (1958) Towards a generalized mereology of Lesniewski, *Studia Logica* , v. 8, pp. 1-154.

Smarandache, F. (2002) A unifying field in logics: neutrosophic logic, *MultipleValued Logic*, v. 8, No. 3, pp. 385-438.

Smith, B. (1996) Mereotopology: A Theory of Parts and Boundaries, *Data and Knowledge Engineering*, v. 20, pp. 287-303.

Smith, B. (1997) Boundaries: an essay in mereotopology, in *The Philosophy of Roderick Chisholm* (Library of Living Philosophers), Open Court, LaSalle, pp. 534–561.

Smith, D. W. (2002) Mathematical Form in the World. *Philosophia Mathematica* 10: ??.

Sneed, J. D. *The Logical Structure of Mathematical Physics*, D. Reidel, Dordrecht, 1979.

Sollins, K.R. (2003) *Effective Design of Naming Systems for Networks*, (electronic edition: http://krs.lcs.mit.edu/regions/docs/naming.pdf).

Stachel, J. (2006) Structure, individuality, and quantum gravity, in *The Structural Foundations of Quantum Gravity*, Oxford University Press, Oxford, pp. 53-82

Stahl, G. E. *Philosophical Principles of Universal Chemistry*, London., 1730.

Steedman, M. (2000) Information Structure and the Syntax-Phonology Interface, *Linguistic Inquiry*, v. 31, pp. 649-689

Steen, L. A. (1988) The Science of Patterns, *Science* , v. 240, No. 4852, pp. 611-616.

Stefanik, R. *The Structures of Reality: Truth, Invariance and Scientific Objectivity*, MSG Press, Washington, 1994.

Stefanik, R. *Structuralism, Category Theory and Philosophy of Mathematics*, MSG Press, Washington, 1994.

Stegmüller, W. *The Structure and Dynamics of Theories*, Springer-Verlag, New York/Heidelberg/Berlin, 1976.

Stegmüller, W. *The Structuralist View. A Possible Analogue of the Bourbaki Programme in Physical Science*, Springer-Verlag, New York/Heidelberg/Berlin, 1979.

Stewart, J. *Calculus: Early Transcendentals*, Brooks/Cole P.C., Pacific Grove, California, 2003

Suppe, F. (Ed.) *The Structure of Scientific Theories,* University of Illinois Press, Urbana, 1974; 1979.

Suppe, F. (1998) The Structure of a Scientific Paper, *Philosophy of Science*, v. 65, pp. 381-405.

Swoyer, C. (1982) The Nature of Natural Laws, *Austral. J. Phil.* , v. 60, pp. 203-223.

Tall, D. O. (2004) The three worlds of mathematics, *For the Learning of Mathematics*, v. 23, No. 3, pp. 29–33.

Tall, D. O. (2004a) Thinking Through Three Worlds of Mathematics, in *Proceedings of the 28th Conference of the International Group for the Psychology of Mathematics Education*, Bergen, Norway, v. 4, pp. 281–288.

Tamassia, R. (1996) Data Structures, *ACM Computing Surveys*, v. 28, No.1, pp. 23-26.

Taschek, W. W. (1995) Belief, Substitution, and Logical Structure, *Nous*, v. 19, pp. 71-95.

Tatievskaia, E. *Russell on the Structure of Propositions*, The Paideia Archive, 1999 (http://www.bu.edu/wcp/).

Tegmark, M. (2008) The Mathematical Universe, *Foundations of Physics* , v. 38, No. 2, pp. 101–150.

Thompson, P. *The structure of biological theories*, SUNY Series in Philosophy and Biology, SUNY Press, 1989.

Tomioka, S. (2007) Information Structure as Information-based Partition, *Interdisciplinary Studies on Information Structure*, v. 6, pp. 97–107.

Tondl, L. (1998) What is the Thematic Structure of Science? *Journal for General Philosophy of Science, Zeitschrift für allgemeine Wissenschaftstheorie*, v. 29, pp. 245-264.

van der Pluijm, B.A. and Marshak, S. *Earth Structure - An Introduction to Structural Geology and Tectonics*, W. W. Norton, New York, 2004.

Varzi, A. C. (1994) On the boundary between mereology and topology, in *Philosophy and Cognitive Science, Proceedings of the 16th International Wittgenstein Symposium*, Hölder-Pichler-Tempsky, Vienna, pp. 419-438.

Varzi, A. C. (2007) *Spatial Reasoning and Ontology: Parts, Wholes, and Locations*, in *Handbook of Spatial Logics*, Springer-Verlag, New York, pp. 945-1038.

von Bayer, H.C. *Information: The New Language of Science*, Harvard University Press, Harvard, 2004.

Warner, M. (2000) The Structure of Imagination, in *Structure in Science and Art*, Cambridge University Press, Cambridge, pp. 33-47.

Wegel, H. (1956) Axiomatische Mengenlehre ohne Elemente von Mengen, *Math. Annalen*, v. 131, pp. 435-462.

von Weizsäcker, C.F. (1958) Die Quantentheorie der einfachen Alternative (Komplementarität und Logik, II), *Zeitschrift für Naturforschung*, v. 13, pp. 245-253

von Weizsäcker, C.F. *Die Einheit der Natur*, Deutscher Taschenbuch Verlag, Munich, Germany, 1974

von Weizsäcker, C.F. *Aufbau der Physik*, Hanser, Munich, Germany, 1985 (Eglish translation: *The Structure of Physics*, Springer, Berlin/Heidelberg/New York, 2006)

Wheeler, J.A. (1990) Information, Physics, Quantum: The Search for Links, in *Complexity, Entropy, and the Physics of Information* (Zurek, W., ed.), Redwood City, CA: Addison-Wesley, pp. 3–28

Wheeler, J. A. *At Home in the Universe*, American Institute of Physics, New York, 1994

Whitehead, A.N. *An Enquiry Concerning the Principles of Natural Knowledge*, Cambridge Univ. Press, 1919.

Whitehead, A.N. *The Concept of Nature*, Cambridge University Press, Cambridge, 1920.

Whyte, L. L. (1955) Note on the Structural Philosophy of Organism, *The British Journal for the Philosophy of Science*, v. 5, No. 20, pp. 332-334.

Wigner, E. (1960) The Unreasonable Effectiveness of Mathematics, *Communications in Pure and Applied Mathematics*, v. 13, pp. 1-14.

Williams, M.S. and Todd, D. *Structures: Theory and Analysis*, Palgrave Macmillan, London, 1999.

Wilimzig, C., Ragert, P. and Dinse H.R. (2012) Cortical topography of intracortical inhibition influences the speed of decision making, *PNAS*, v.109, No. 8, pp. 3107-3112.

Williams, B. (1986) Hylomorphism, *Oxford Studies in Ancient Philosophy*, v. 4, pp. 189–199

Wirth, N. *Algorithms + Data Structures = Programs*, Prentice Hall, Englewood Cliffs, NJ, 1975.

Wirth, N. *Algorithms and Data Structures*, Prentice Hall, Englewood Cliffs, NJ, 1986.

Worrall, J. (1989) Structural Realism: The Best of Both Worlds? *Dialectica*, v. 43, pp. 99-124.

Worrall, J. (2007) Miracles and Models: Why reports of the death of Structural Realism may be exaggerated, *Royal Institute of Philosophy*, v. 82, Supplement, pp. 125-154.

Wulftange, J. and Greene, M. *Hylomorphism and contemporary physics*, Woodstock College Press, Woodstock, MD, 1952

Wyatt, C. S. *The Existential Primer* (http://www.tameri.com/csw/exist/hegel.shtml).

Yaglom, I.M. *Mathematical Structures and Mathematical Modeling*, Sov. Radio, Moscow, 1980 (in Russian).

Young, F.W. *Structural Ecology of Health and Community*, The Internet-First University Press, Cornell University, Ithaca, New York, 2009

Young, J., Sharlin, E. and Igarashi, T. (2011) What Is Mixed Reality, Anyway? Considering the Boundaries of Mixed Reality in the Context of Robots, in Mixed *Reality and Human-Robot Interaction*, ISCA 47, pp. 1–11.

Yourdon, E. *Managing the Structured Techniques*, Prentice-Hall, Englewood Cliffs, NJ, 1979.

Zadeh, L. (1965) Fuzzy Sets, *Information and Control*, v. 8, No. 3, pp. 338-353.

Zahar, E. G. (1996) Poincaré's Structural Realism and his Logic of Discovery, in *Henri Poincaré: Science and Philosophy*, Akademie Verlag & A. Blanchard, Berlin, pp. 45-68.

Zahar, E. *Poincaré's Philosophy: From Conventionalism to Phenomenology*, Open Court, Chicago and La Salle, 2001.

Zalta, E. *Abstract Objects: An Introduction to Axiomatic Metaphysics*, D. Reidel, Dordrecht, 1983.

Zhao, Q. (2011) 10 Scientific Problems in Virtual Reality, *Communications of the ACM*, v. 54, No. 2, pp. 116-118.

Zimmel, G. *The Sociology of Georg Zimmel*, The Free Press, New York, 1964

INDEX

#

A

B

C

D

H

I

J

K

L

M

N

O

P

Q

T

U

V

W

Y

formati